PSSA Subject Test

Mathematics Grade 7

Student Practice Workbook

+ Two Full-Length PSSA Math Tests

Math Notion

www.MathNotion.com

PSSA Subject Test Mathematics Grade 7

PSSA Subject Test Mathematics Grade 7

Published in the United State of America By

The Math Notion

Web: WWW.MathNotion.com

Email: info@Mathnotion.com

Copyright © 2021 by the Math Notion. All rights reserved. No part of this publication may be reproduced, stored in a retrieval system, or transmitted in any form or by any means, electronic, mechanical, photocopying, recording, scanning, or otherwise, except as permitted under Section 107 or 108 of the 1976 United States Copyright Ac, without permission of the author.

All inquiries should be addressed to the Math Notion.

ISBN: 978-1-63620-067-5

The Math Notion

Michael Smith has been a math instructor for over a decade now. He launched the Math Notion. Since 2006, we have devoted our time to both teaching and developing exceptional math learning materials. As a test prep company, we have worked with thousands of students. We have used the feedback of our students to develop a unique study program that can be used by students to drastically improve their math scores fast and effectively. We have more than a thousand Math learning books including:

- **SAT Math Prep**
- **ACT Math Prep**
- **SSAT/ISEE Math Prep**
- **Accuplacer Math Prep**
- **Common Core Math Prep**
- **many Math Education Workbooks, Study Guides, Practice and Exercise Books**

As an experienced Math test preparation company, we have helped many students raise their standardized test scores—and attend the colleges of their dreams: We tutor online and in person, we teach students in large groups, and we provide training materials and textbooks through our website and through Amazon.

You can contact us via email at:

info@Mathnotion.com

PSSA Subject Test Mathematics Grade 7

Get the Targeted Practice You Need to Ace the PSSA Math Test!

PSSA Subject Test Mathematics Grade 7 includes easy-to-follow instructions, helpful examples, and plenty of math practice problems to assist students to master each concept, brush up their problem-solving skills, and create confidence.

The PSSA math practice book provides numerous opportunities to evaluate basic skills along with abundant remediation and intervention activities. It is a skill that permits you to quickly master intricate information and produce better leads in less time.

Students can boost their test-taking skills by taking the book's two practice PSSA Math exams. All test questions answered and explained in detail.

Important Features of the 7th grade PSSA Math Book:

- A **complete review** of PSSA math test topics,
- Over 2,500 practice problems covering all topics tested,
- The most important concepts you need to know,
- Clear and concise, easy-to-follow sections,
- Well designed for enhanced learning and interest,
- Hands-on experience with all question types,
- **2 full-length practice tests** with detailed answer explanations,
- Cost-Effective Pricing,

Powerful math exercises to help you avoid traps and pacing yourself to beat the PSSA Pennsylvania test. Students will gain valuable experience and raise their confidence by taking 7th grade math practice tests, learning about test structure, and gaining a deeper understanding of what is tested on the PSSA math grade 7. If ever there was a book to respond to the pressure to increase students' test scores, this is it.

WWW.MathNotion.COM

… So Much More Online!

- ✓ FREE Math Lessons
- ✓ More Math Learning Books!
- ✓ Mathematics Worksheets
- ✓ Online Math Tutors

For a PDF Version of This Book

Please Visit WWW.MathNotion.com

PSSA Subject Test Mathematics Grade 7

Contents

Chapter 1 : Integers and Number Theory **11**
   Rounding 12
   Rounding and Estimates 13
   Adding and Subtracting Integers 14
   Multiplying and Dividing Integers 15
   Order of Operations 16
   Ordering Integers and Numbers 17
   Integers and Absolute Value 18
   Factoring Numbers 19
   Greatest Common Factor 20
   Least Common Multiple 21
   Answers of Worksheets 22

Chapter 2 : Fractions **25**
   Simplifying Fractions 26
   Adding and Subtracting Fractions 27
   Multiplying and Dividing Fractions 28
   Adding and Subtracting Mixed Numbers 29
   Multiplying and Dividing Mixed Numbers 30
   Answers of Worksheets 31

Chapter 3 : Decimals **33**
   Adding and Subtracting Decimals 34
   Multiplying and Dividing Decimals 35
   Comparing Decimals 36
   Rounding Decimals 37
   Answers of Worksheets 38

Chapter 4 : Proportions, Ratios, and Percent **39**
   Simplifying Ratios 40
   Proportional Ratios 41
   Similarity and Ratios 42
   Ratio and Rates Word Problems 43
   Percentage Calculations 44
   Percent Problems 45
   Discount, Tax and Tip 46

WWW.MathNotion.Com

PSSA Subject Test Mathematics Grade 7

Percent of Change 47
Simple Interest 48
Answers of Worksheets 49

Chapter 5 : Exponents and Radicals Expressions 52
Multiplication Property of Exponents 53
Zero and Negative Exponents 54
Division Property of Exponents 55
Powers of Products and Quotients 56
Negative Exponents and Negative Bases 57
Scientific Notation 58
Square Roots 59
Answers of Worksheets 60

Chapter 6 : Algebraic Expressions 63
Translate Phrases into an Algebraic Statement 64
Simplifying Variable Expressions 65
The Distributive Property 66
Evaluating One Variable Expressions 67
Evaluating Two Variables Expressions 68
Combining like Terms 69
Answers of Worksheets 70

Chapter 7 : Equations and Inequalities 72
One–Step Equations 73
Multi–Step Equations 74
Graphing Single–Variable Inequalities 75
One–Step Inequalities 76
Multi-Step Inequalities 77
Systems of Equations 78
Systems of Equations Word Problems 79
Answers of Worksheets 80

Chapter 8 : Linear Functions 83
Finding Slope 84
Graphing Lines Using Line Equation 85
Writing Linear Equations 86
Graphing Linear Inequalities 87
Finding Midpoint 88
Finding Distance of Two Points 89
Answers of Worksheets 90

PSSA Subject Test Mathematics Grade 7

Chapter 9 : Transformations 93
    Translations 94
    Reflections 95
    Rotations 97
    Dilations 99
    Coordinates of Vertices 100
    Answers of Worksheets 101

Chapter 10 : Geometry and Solid Figures 104
    Angles.. 105
    Pythagorean Relationship 106
    Triangles.. 107
    Polygons 108
    Trapezoids 109
    Circles 110
    Cubes 111
    Rectangular Prism 111
    Cylinder.. 113
    Pyramids and Cone.. 114
    Answers of Worksheets 115

Chapter 11 : Statistics and Probability 117
    Mean and Median 118
    Mode and Range 119
    Times Series 120
    Stem-and-Leaf Plot.. 121
    Pie Graph 122
    Probability Problems 123
    Answers of Worksheets 124

Chapter 12 : PSSA Math Practice Tests 127
    PSSA Practice Test 1... 131
    PSSA Practice Test 2... 139

Chapter 13 : Answers and Explanations.. 147
    Answer Key.. 147
    Practice Test 1 149
    Practice Test 2 153

PSSA Subject Test Mathematics Grade 7

Chapter 1 :
Integers and Number Theory

Topics that you will practice in this chapter:

- ✓ Rounding
- ✓ Rounding and Estimates
- ✓ Adding and Subtracting Integers
- ✓ Multiplying and Dividing Integers
- ✓ Order of Operations
- ✓ Ordering Integers and Numbers
- ✓ Integers and Absolute Value
- ✓ Factoring Numbers
- ✓ Greatest Common Factor (GCF)
- ✓ Least Common Multiple (LCM)

"Wherever there is number, there is beauty." –Proclus

Rounding

✏️ **Round each number to the nearest ten.**

1) 42 = ____ 5) 19 = ____ 9) 48 = ____

2) 88 = ____ 6) 25 = ____ 10) 81 = ____

3) 24 = ____ 7) 93 = ____ 11) 58 = ____

4) 57 = ____ 8) 71 = ____ 12) 87 = ____

✏️ **Round each number to the nearest hundred.**

13) 198 = ____ 17) 321 = ____ 21) 580 = ____

14) 387 = ____ 18) 433 = ____ 22) 868 = ____

15) 816 = ____ 19) 579 = ____ 23) 480 = ____

16) 101 = ____ 20) 825 = ____ 24) 287 = ____

✏️ **Round each number to the nearest thousand.**

25) 1,382 = ____ 29) 9,099 = ____ 33) 52,866 = ____

26) 3,420 = ____ 30) 22,980 = ____ 34) 85,190 = ____

27) 4,254 = ____ 31) 45,188 = ____ 35) 70,990 = ____

28) 6,861 = ____ 32) 16,808 = ____ 36) 26,869 = ____

PSSA Subject Test Mathematics Grade 7

Rounding and Estimates

✎ **Estimate the sum by rounding each number to the nearest ten.**

1) 13 + 22 = _____

2) 71 + 23 = _____

3) 61 + 58 = _____

4) 56 + 85 = _____

5) 368 + 249 = _____

6) 330 + 903 = _____

7) 471 + 293 = _____

8) 1,950 + 2,655 = _____

✎ **Estimate the product by rounding each number to the nearest ten.**

9) 32 × 71 = _____

10) 12 × 33 = _____

11) 31 × 83 = _____

12) 19 × 11 = _____

13) 42 × 76 = _____

14) 63 × 34 = _____

15) 19 × 31 = _____

16) 59 × 71 = _____

✎ **Estimate the sum or product by rounding each number to the nearest ten.**

17) 29 × 12

18) 37 × 26

19) 48 + 82

20) 65 + 44

21) 37 × 14

22) 71 + 32

WWW.MathNotion.Com

Adding and Subtracting Integers

✏️ **Find each sum.**

1) $14 + (-6) =$

2) $(-13) + (-20) =$

3) $5 + (-28) =$

4) $50 + (-12) =$

5) $(-7) + (-15) + 3 =$

6) $30 + (-14) + 8 =$

7) $40 + (-10) + (-14) + 17 =$

8) $(-15) + (-20) + 13 + 35 =$

9) $40 + (-20) + (38 - 29) =$

10) $28 + (-12) + (30 - 12) =$

✏️ **Find each difference.**

11) $(-18) - (-7) =$

12) $25 - (-14) =$

13) $(-20) - 36 =$

14) $34 - (-19) =$

15) $51 - (30 - 21) =$

16) $17 - (5) - (-24) =$

17) $(35 + 20) - (-46) =$

18) $48 - 16 - (-8) =$

19) $62 - (28 + 17) - (-15) =$

20) $58 - (-23) - (-31) =$

21) $19 - (-8) - (-13) =$

22) $(19 - 24) - (-14) =$

23) $27 - 33 - (-21) =$

24) $58 - (32 + 24) - (-9) =$

25) $36 - (-30) + (-17) =$

26) $27 - (-42) + (-31) =$

PSSA Subject Test Mathematics Grade 7

Multiplying and Dividing Integers

✎ **Find each product.**

1) $(-9) \times (-5) =$

2) $(-3) \times 9 =$

3) $8 \times (-12) =$

4) $(-7) \times (-20) =$

5) $(-3) \times (-5) \times 6 =$

6) $(14 - 3) \times (-8) =$

7) $12 \times (-9) \times (-3) =$

8) $(140 + 10) \times (-2) =$

9) $10 \times (-12 + 8) \times 3 =$

10) $(-8) \times (-5) \times (-10) =$

✎ **Find each quotient.**

11) $42 \div (-7) =$

12) $(-48) \div (-6) =$

13) $(-40) \div (-8) =$

14) $54 \div (-2) =$

15) $152 \div 19 =$

16) $(-144) \div (-12) =$

17) $180 \div (-10) =$

18) $(-312) \div (-12) =$

19) $221 \div (-13) =$

20) $(-126) \div (6) =$

21) $(-161) \div (-7) =$

22) $-266 \div (-14) =$

23) $(-120) \div (-4) =$

24) $270 \div (-18) =$

25) $(-208) \div (-8) =$

26) $(135) \div (-15) =$

Order of Operations

✏️ **Evaluate each expression.**

1) $7 + (5 \times 4) =$

2) $14 - (3 \times 6) =$

3) $(19 \times 4) + 16 =$

4) $(16 - 7) - (8 \times 2) =$

5) $27 + (18 \div 3) =$

6) $(18 \times 8) \div 6 =$

7) $(32 \div 4) \times (-2) =$

8) $(9 \times 4) + (32 - 18) =$

9) $24 + (4 \times 3) + 7 =$

10) $(36 \times 3) \div (2 + 2) =$

11) $(-7) + (12 \times 3) + 11 =$

12) $(8 \times 5) - (24 \div 6) =$

13) $(7 \times 6 \div 3) - (12 + 9) =$

14) $(13 + 5 - 14) \times 3 - 2 =$

15) $(20 - 14 + 30) \times (64 \div 4) =$

16) $32 + (28 - (36 \div 9)) =$

17) $(7 + 6 - 4 - 7) + (15 \div 5) =$

18) $(85 - 20) + (20 - 18 + 7) =$

19) $(20 \times 2) + (14 \times 3) - 22 =$

20) $18 + 5 - (30 \times 3) + 20 =$

PSSA Subject Test Mathematics Grade 7

Ordering Integers and Numbers

✏ **Order each set of integers from least to greatest.**

1) $8, -10, -5, -3, 4$ ___, ___, ___, ___, ___, ___

2) $-10, -18, 6, 14, 27$ ___, ___, ___, ___, ___, ___

3) $15, -8, -21, 21, -23$ ___, ___, ___, ___, ___, ___

4) $-14, -40, 23, -12, 47$ ___, ___, ___, ___, ___, ___

5) $59, -54, 32, -57, 36$ ___, ___, ___, ___, ___, ___

6) $68, 26, -19, 47, -34$ ___, ___, ___, ___, ___, ___

✏ **Order each set of integers from greatest to least.**

7) $18, 36, -16, -18, -10$ ___, ___, ___, ___, ___, ___

8) $27, 34, -12, -24, 94$ ___, ___, ___, ___, ___, ___

9) $50, -21, -13, 42, -2$ ___, ___, ___, ___, ___, ___

10) $37, 46, -20, -16, 86$ ___, ___, ___, ___, ___, ___

11) $-18, 88, -26, -59, 75$ ___, ___, ___, ___, ___, ___

12) $-65, -30, -25, 3, 14$ ___, ___, ___, ___, ___, ___

WWW.MathNotion.Com

PSSA Subject Test Mathematics Grade 7

Integers and Absolute Value

✎ **Write absolute value of each number.**

1) $|-2| =$

2) $|-27| =$

3) $|-20| =$

4) $|14| =$

5) $|6| =$

6) $|-55| =$

7) $|16| =$

8) $|2| =$

9) $|54| =$

10) $|-4| =$

11) $|-11|$

12) $|88| =$

13) $|0| =$

14) $|79| =$

15) $|-32| =$

16) $|-17| =$

17) $|42| =$

18) $|-46| =$

19) $|1| =$

20) $|-40| =$

✎ **Evaluate the value.**

21) $|-5| - \frac{|-21|}{7} =$

22) $14 - |3 - 15| - |-4| =$

23) $\frac{|-32|}{4} \times |-4| =$

24) $\frac{|7 \times (-3)|}{7} \times \frac{|-19|}{3} =$

25) $|4 \times (-5)| + \frac{|-40|}{5} =$

26) $\frac{|-45|}{9} \times \frac{|-24|}{12} =$

27) $|-12 + 8| \times \frac{|-7 \times 7|}{7}$

28) $\frac{|-11 \times 2|}{4} \times |-16| =$

Factoring Numbers

✎ **List all positive factors of each number.**

1) 9

2) 16

3) 24

4) 30

5) 26

6) 46

7) 20

8) 68

9) 28

10) 98

11) 14

12) 54

13) 55

14) 18

15) 63

16) 34

17) 50

18) 62

19) 95

20) 64

21) 70

22) 45

23) 22

24) 65

PSSA Subject Test Mathematics Grade 7

Greatest Common Factor

✏ **Find the GCF for each number pair.**

1) 6, 2

2) 4, 5

3) 3, 12

4) 7, 3

5) 5, 10

6) 8, 48

7) 6, 18

8) 9, 15

9) 12, 18

10) 4, 36

11) 6, 10

12) 28, 52

13) 25, 10

14) 22, 24

15) 9, 54

16) 8, 54

17) 42, 14

18) 16, 40

19) 9, 2, 3

20) 5, 15, 10

21) 7, 9, 2

22) 16, 64

23) 30, 48

24) 36, 63

PSSA Subject Test Mathematics Grade 7

Least Common Multiple

✏ Find the LCM for each number pair.

1) 6, 9

2) 15, 45

3) 16, 40

4) 12, 36

5) 18, 27

6) 14, 42

7) 6, 30

8) 8, 56

9) 7, 21

10) 8, 20

11) 15, 25

12) 7, 9

13) 4, 11

14) 8, 28

15) 28, 56

16) 40, 50

17) 12, 13

18) 22, 11

19) 36, 20

20) 15, 35

21) 18, 81

22) 30, 54

23) 18, 45

24) 75, 25

WWW.MathNotion.Com

PSSA Subject Test Mathematics Grade 7

Answers of Worksheets

Rounding

1) 40	10) 80	19) 600	28) 7,000
2) 90	11) 60	20) 800	29) 9,000
3) 20	12) 90	21) 600	30) 23,000
4) 60	13) 200	22) 900	31) 45,000
5) 20	14) 400	23) 500	32) 17,000
6) 30	15) 800	24) 300	33) 53,000
7) 90	16) 100	25) 1,000	34) 85,000
8) 70	17) 300	26) 3,000	35) 71,000
9) 50	18) 400	27) 4,000	36) 27,000

Rounding and Estimates

1) 30	7) 760	13) 3,200	19) 130
2) 90	8) 4,610	14) 1,800	20) 110
3) 120	9) 2,100	15) 600	21) 400
4) 150	10) 300	16) 4,200	22) 100
5) 620	11) 2,400	17) 300	
6) 1,230	12) 200	18) 1,200	

Adding and Subtracting Integers

1) 8	8) 13	15) 42	22) 9
2) −33	9) 29	16) 36	23) 15
3) −23	10) 34	17) 101	24) 11
4) 38	11) −11	18) 40	25) 49
5) −19	12) 39	19) 32	26) 38
6) 24	13) −56	20) 112	
7) 33	14) 53	21) 40	

Multiplying and Dividing Integers

1) 45	6) −88	11) −6	16) 12
2) −27	7) 324	12) 8	17) −18
3) −96	8) −300	13) 5	18) 26
4) 140	9) −120	14) −27	19) −17
5) 90	10) −400	15) 8	20) −21

WWW.MathNotion.Com

PSSA Subject Test Mathematics Grade 7

21) 23   23) 30   25) 26
22) 19   24) −15   26) −9

Order of Operations

1) 27   6) 24   11) 40   16) 56
2) −4   7) −16   12) 36   17) 5
3) 92   8) 50   13) −7   18) 74
4) −7   9) 43   14) 10   19) 60
5) 33   10) 27   15) 576   20) −47

Ordering Integers and Numbers

1) −10, −5, −3, 4, 8   7) 36, 18, −10, −16, −18
2) −18, −10, 6, 14, 27   8) 94, 34, 27, −12, −24
3) −23, −21, −8, 15, 21   9) 50, 42, −2, −13, −21
4) −40, −14, −12, 23, 47   10) 86, 46, 37, −16, −20
5) −57, −54, 32, 36, 59   11) 88, 75, −18, −26, −59
6) −34, −19, 26, 47, 68   12) 14, 3, −25, −30, −65

Integers and Absolute Value

1) 2   8) 2   15) 32   22) −2
2) 27   9) 54   16) 17   23) 32
3) 20   10) 4   17) 42   24) 19
4) 14   11) 11   18) 46   25) 28
5) 6   12) 88   19) 1   26) 10
6) 55   13) 0   20) 40   27) 28
7) 16   14) 79   21) 2   28) 88

Factoring Numbers

1) 1, 3, 9   9) 1, 2, 4, 7, 14, 28   17) 1, 2, 5, 10, 25, 50
2) 1, 2, 4, 8, 16   10) 1, 2, 7, 14, 49, 98   18) 1, 2, 31, 62
3) 1, 2, 3, 4, 6, 8, 12, 24   11) 1, 2, 7, 14   19) 1, 5, 19, 95
4) 1, 2, 3, 5, 6, 10, 15, 30   12) 1, 2, 3, 6, 9, 18, 27, 54 20) 1, 2, 4, 8, 16, 32, 64
5) 1, 2, 13, 26   13) 1, 5, 11, 55   21) 1, 2, 5, 7, 10, 14, 35, 70
6) 1, 2, 23, 46   14) 1, 2, 3, 6, 9, 18   22) 1, 3, 5, 9, 15, 45
7) 1, 2, 4, 5, 10, 20   15) 1, 3, 7, 9, 21, 63   23) 1, 2, 11, 22
8) 1, 2, 4, 17, 34, 68   16) 1, 2, 17, 34   24) 1, 5, 13, 65

PSSA Subject Test Mathematics Grade 7

Greatest Common Factor

1) 2
2) 1
3) 3
4) 1
5) 5
6) 8
7) 6
8) 3
9) 6
10) 4
11) 2
12) 4
13) 5
14) 2
15) 9
16) 2
17) 14
18) 8
19) 1
20) 5
21) 1
22) 16
23) 6
24) 9

Least Common Multiple

1) 18
2) 45
3) 80
4) 36
5) 54
6) 42
7) 30
8) 56
9) 21
10) 40
11) 75
12) 63
13) 44
14) 56
15) 56
16) 200
17) 156
18) 22
19) 180
20) 105
21) 162
22) 270
23) 90
24) 75

Chapter 2: Fractions

Topics that you will practice in this chapter:

- ✓ Simplifying Fractions
- ✓ Adding and Subtracting Fractions
- ✓ Multiplying and Dividing Fractions
- ✓ Adding and Subtract Mixed Numbers
- ✓ Multiplying and Dividing Mixed Numbers

"A Man is like a fraction whose numerator is what he is and whose denominator is what he thinks of himself. The larger the denominator, the smaller the fraction." –Tolstoy

PSSA Subject Test Mathematics Grade 7

Simplifying Fractions

✎ Simplify each fraction to its lowest terms.

1) $\frac{5}{10} =$

2) $\frac{28}{35} =$

3) $\frac{27}{36} =$

4) $\frac{40}{80} =$

5) $\frac{14}{56} =$

6) $\frac{32}{48} =$

7) $\frac{52}{65} =$

8) $\frac{15}{60} =$

9) $\frac{80}{160} =$

10) $\frac{55}{77} =$

11) $\frac{28}{112} =$

12) $\frac{32}{64} =$

13) $\frac{63}{72} =$

14) $\frac{81}{90} =$

15) $\frac{35}{105} =$

16) $\frac{25}{70} =$

17) $\frac{80}{280} =$

18) $\frac{12}{81} =$

19) $\frac{36}{186} =$

20) $\frac{240}{540} =$

21) $\frac{70}{560} =$

✎ Find the answer for each problem.

22) Which of the following fractions equal to $\frac{3}{4}$? ____

　　A. $\frac{60}{90}$ 　　B. $\frac{43}{104}$　　C. $\frac{48}{64}$　　D. $\frac{150}{300}$

23) Which of the following fractions equal to $\frac{5}{8}$? ____

　　A. $\frac{125}{200}$　　B. $\frac{115}{200}$　　C. $\frac{50}{100}$　　D. $\frac{30}{90}$

24) Which of the following fractions equal to $\frac{3}{7}$? ____

　　A. $\frac{58}{116}$　　B. $\frac{54}{126}$　　C. $\frac{270}{167}$　　D. $\frac{42}{63}$

WWW.MathNotion.Com

PSSA Subject Test Mathematics Grade 7

Adding and Subtracting Fractions

✎ **Find the sum.**

1) $\frac{5}{9} + \frac{4}{9} =$

2) $\frac{1}{2} + \frac{1}{7} =$

3) $\frac{3}{8} + \frac{1}{4} =$

4) $\frac{3}{5} + \frac{1}{2} =$

5) $\frac{1}{4} + \frac{3}{5} =$

6) $\frac{7}{8} + \frac{3}{8} =$

7) $\frac{1}{2} + \frac{7}{10} =$

8) $\frac{2}{5} + \frac{2}{3} =$

9) $\frac{5}{7} + \frac{2}{3} =$

10) $\frac{7}{12} + \frac{3}{4} =$

11) $\frac{5}{6} + \frac{2}{5} =$

12) $\frac{1}{12} + \frac{2}{3} =$

✎ **Find the difference.**

13) $\frac{1}{3} - \frac{1}{6} =$

14) $\frac{3}{4} - \frac{1}{8} =$

15) $\frac{1}{2} - \frac{1}{3} =$

16) $\frac{1}{4} - \frac{1}{5} =$

17) $\frac{5}{8} - \frac{2}{3} =$

18) $\frac{1}{4} - \frac{1}{7} =$

19) $\frac{5}{6} - \frac{1}{9} =$

20) $\frac{3}{4} - \frac{1}{6} =$

21) $\frac{7}{8} - \frac{1}{12} =$

22) $\frac{8}{15} - \frac{3}{5} =$

23) $\frac{3}{12} - \frac{1}{14} =$

24) $\frac{10}{13} - \frac{7}{26} =$

25) $\frac{6}{7} - \frac{3}{4} =$

26) $\frac{4}{5} - \frac{1}{8} =$

27) $\frac{4}{7} - \frac{2}{35} =$

28) $\frac{9}{16} - \frac{2}{8} =$

29) $\frac{8}{9} - \frac{7}{18} =$

30) $\frac{1}{2} - \frac{4}{9} =$

WWW.MathNotion.Com

PSSA Subject Test Mathematics Grade 7

Multiplying and Dividing Fractions

✏ Find the value of each expression in lowest terms.

1) $\frac{1}{5} \times \frac{15}{5} =$

2) $\frac{9}{12} \times \frac{4}{9} =$

3) $\frac{1}{16} \times \frac{8}{10} =$

4) $\frac{1}{24} \times \frac{8}{10} =$

5) $\frac{1}{5} \times \frac{1}{4} =$

6) $\frac{7}{9} \times \frac{1}{7} =$

7) $\frac{6}{7} \times \frac{1}{3} =$

8) $\frac{2}{8} \times \frac{2}{8} =$

9) $\frac{5}{8} \times \frac{3}{5} =$

10) $\frac{4}{7} \times \frac{1}{8} =$

11) $\frac{7}{15} \times \frac{5}{7} =$

12) $\frac{3}{10} \times \frac{5}{9} =$

✏ Find the value of each expression in lowest terms.

13) $\frac{1}{4} \div \frac{1}{8} =$

14) $\frac{1}{10} \div \frac{1}{5} =$

15) $\frac{3}{4} \div \frac{1}{5} =$

16) $\frac{1}{3} \div \frac{5}{6} =$

17) $\frac{1}{7} \div \frac{8}{42} =$

18) $\frac{3}{4} \div \frac{1}{6} =$

19) $\frac{2}{7} \div \frac{7}{13} =$

20) $\frac{1}{24} \div \frac{3}{16} =$

21) $\frac{7}{12} \div \frac{5}{6} =$

22) $\frac{22}{18} \div \frac{11}{9} =$

23) $\frac{9}{35} \div \frac{3}{7} =$

24) $\frac{2}{7} \div \frac{8}{21} =$

25) $\frac{1}{9} \div \frac{2}{5} =$

26) $\frac{5}{12} \div \frac{3}{5} =$

27) $\frac{3}{20} \div \frac{1}{6} =$

28) $\frac{8}{20} \div \frac{3}{4} =$

29) $\frac{5}{6} \div \frac{2}{9} =$

30) $\frac{5}{11} \div \frac{3}{4} =$

WWW.MathNotion.Com

PSSA Subject Test Mathematics Grade 7

Adding and Subtracting Mixed Numbers

✏ **Find the sum.**

1) $3\frac{1}{3} + 2\frac{1}{6} =$

2) $4\frac{1}{2} + 3\frac{1}{2} =$

3) $3\frac{3}{8} + 1\frac{1}{8} =$

4) $2\frac{1}{4} + 2\frac{1}{3} =$

5) $3\frac{5}{6} + 2\frac{7}{12} =$

6) $5\frac{4}{15} + 3\frac{3}{5} =$

7) $2\frac{1}{3} + 4\frac{3}{7} =$

8) $3\frac{1}{2} + 4\frac{2}{5} =$

9) $5\frac{2}{5} + 6\frac{3}{7} =$

10) $8\frac{5}{16} + 6\frac{1}{12} =$

✏ **Find the difference.**

11) $3\frac{1}{4} - 1\frac{3}{4} =$

12) $6\frac{3}{5} - 4\frac{2}{5} =$

13) $4\frac{1}{3} - 3\frac{1}{9} =$

14) $7\frac{1}{7} - 5\frac{1}{2} =$

15) $5\frac{1}{3} - 2\frac{1}{12} =$

16) $8\frac{1}{5} - 4\frac{1}{3} =$

17) $9\frac{1}{4} - 6\frac{1}{8} =$

18) $11\frac{7}{15} - 8\frac{3}{5} =$

19) $14\frac{5}{6} - 11\frac{3}{5} =$

20) $18\frac{2}{7} - 14\frac{1}{5} =$

21) $9\frac{1}{3} - 4\frac{1}{4} =$

22) $6\frac{1}{8} - 4\frac{1}{16} =$

23) $19\frac{3}{8} - 15\frac{1}{3} =$

24) $11\frac{1}{9} - 8\frac{1}{8} =$

25) $17\frac{1}{7} - 11\frac{1}{5} =$

26) $16\frac{2}{9} - 9\frac{5}{7} =$

WWW.MathNotion.Com

PSSA Subject Test Mathematics Grade 7

Multiplying and Dividing Mixed Numbers

✎ **Find the product.**

1) $5\frac{1}{2} \times 2\frac{1}{4} =$

2) $5\frac{1}{3} \times 4\frac{1}{3} =$

3) $5\frac{3}{4} \times 6\frac{1}{4} =$

4) $3\frac{1}{3} \times 2\frac{3}{5} =$

5) $4\frac{8}{10} \times 1\frac{1}{24} =$

6) $6\frac{2}{7} \times 1\frac{1}{11} =$

7) $8\frac{2}{3} \times 3\frac{1}{2} =$

8) $3\frac{4}{7} \times 2\frac{1}{5} =$

9) $5\frac{2}{8} \times 4\frac{1}{6} =$

10) $7\frac{3}{3} \times 1\frac{3}{8} =$

✎ **Find the quotient.**

11) $2\frac{2}{5} \div 4\frac{1}{5} =$

12) $4\frac{1}{6} \div 3\frac{1}{3} =$

13) $6\frac{1}{3} \div 1\frac{1}{2} =$

14) $7\frac{1}{10} \div 2\frac{2}{5} =$

15) $3\frac{1}{3} \div 1\frac{1}{9} =$

16) $1\frac{1}{10} \div 4\frac{1}{2} =$

17) $1\frac{3}{16} \div 5\frac{1}{4} =$

18) $4\frac{1}{3} \div 4\frac{3}{4} =$

19) $9\frac{1}{3} \div 2\frac{1}{4} =$

20) $15\frac{1}{3} \div 5\frac{1}{2} =$

21) $4\frac{1}{6} \div 1\frac{1}{5} =$

22) $1\frac{1}{18} \div 1\frac{2}{9} =$

23) $4\frac{2}{7} \div 1\frac{3}{10} =$

24) $7\frac{1}{3} \div 2\frac{2}{11} =$

25) $8\frac{2}{5} \div 1\frac{1}{6} =$

26) $9\frac{1}{3} \div 2\frac{1}{7} =$

WWW.MathNotion.Com

PSSA Subject Test Mathematics Grade 7

Answers of Worksheets

Simplifying Fractions

1) $\frac{1}{2}$
2) $\frac{4}{5}$
3) $\frac{3}{4}$
4) $\frac{1}{2}$
5) $\frac{1}{4}$
6) $\frac{2}{3}$
7) $\frac{4}{5}$
8) $\frac{1}{4}$
9) $\frac{1}{2}$
10) $\frac{5}{7}$
11) $\frac{1}{4}$
12) $\frac{1}{2}$
13) $\frac{7}{8}$
14) $\frac{9}{10}$
15) $\frac{1}{3}$
16) $\frac{5}{14}$
17) $\frac{2}{7}$
18) $\frac{4}{27}$
19) $\frac{6}{31}$
20) $\frac{4}{9}$
21) $\frac{1}{8}$
22) C
23) A
24) B

Adding and Subtracting Fractions

1) $\frac{9}{9} = 1$
2) $\frac{9}{14}$
3) $\frac{5}{8}$
4) $1\frac{1}{10}$
5) $\frac{17}{20}$
6) $1\frac{1}{4}$
7) $1\frac{1}{5}$
8) $1\frac{1}{15}$
9) $1\frac{8}{21}$
10) $1\frac{1}{3}$
11) $1\frac{7}{30}$
12) $\frac{3}{4}$
13) $\frac{1}{6}$
14) $\frac{5}{8}$
15) $\frac{1}{6}$
16) $\frac{1}{20}$
17) $-\frac{1}{24}$
18) $\frac{3}{28}$
19) $\frac{13}{18}$
20) $\frac{7}{12}$
21) $\frac{19}{24}$
22) $-\frac{1}{15}$
23) $\frac{5}{28}$
24) $\frac{1}{2}$
25) $\frac{3}{28}$
26) $\frac{27}{40}$
27) $\frac{18}{35}$
28) $\frac{5}{16}$
29) $\frac{1}{2}$
30) $\frac{1}{18}$

Multiplying and Dividing Fractions

1) $\frac{3}{5}$
2) $\frac{1}{3}$
3) $\frac{1}{20}$
4) $\frac{1}{30}$
5) $\frac{1}{20}$
6) $\frac{1}{9}$
7) $\frac{2}{7}$
8) $\frac{1}{16}$
9) $\frac{3}{8}$
10) $\frac{1}{14}$
11) $\frac{1}{3}$
12) $\frac{1}{6}$
13) 2
14) $\frac{1}{2}$
15) $3\frac{3}{4}$
16) $\frac{2}{5}$

PSSA Subject Test Mathematics Grade 7

17) $\frac{3}{4}$

18) $4\frac{1}{2}$

19) $\frac{26}{49}$

20) $\frac{2}{9}$

21) $\frac{7}{10}$

22) 1

23) $\frac{3}{5}$

24) $\frac{3}{4}$

25) $\frac{5}{18}$

26) $\frac{25}{36}$

27) $\frac{9}{10}$

28) $\frac{8}{15}$

29) $3\frac{3}{4}$

30) $\frac{20}{33}$

Adding and Subtracting Mixed Numbers

1) $5\frac{1}{2}$

2) 8

3) $4\frac{1}{2}$

4) $4\frac{7}{12}$

5) $6\frac{5}{12}$

6) $8\frac{13}{15}$

7) $6\frac{16}{21}$

8) $7\frac{9}{10}$

9) $11\frac{29}{35}$

10) $14\frac{19}{48}$

11) $1\frac{1}{2}$

12) $2\frac{1}{5}$

13) $1\frac{2}{9}$

14) $1\frac{9}{14}$

15) $3\frac{1}{4}$

16) $3\frac{13}{15}$

17) $3\frac{1}{8}$

18) $2\frac{13}{15}$

19) $3\frac{7}{30}$

20) $4\frac{3}{35}$

21) $5\frac{1}{12}$

22) $2\frac{1}{16}$

23) $4\frac{1}{24}$

24) $2\frac{71}{72}$

25) $5\frac{33}{35}$

26) $6\frac{32}{63}$

Multiplying and Dividing Mixed Numbers

1) $12\frac{3}{8}$

2) $23\frac{1}{9}$

3) $35\frac{15}{16}$

4) $8\frac{2}{3}$

5) 5

6) $6\frac{6}{7}$

7) $30\frac{1}{3}$

8) $7\frac{6}{7}$

9) $21\frac{7}{8}$

10) 11

11) $\frac{4}{7}$

12) $1\frac{1}{4}$

13) $4\frac{2}{9}$

14) $2\frac{23}{24}$

15) 3

16) $\frac{11}{45}$

17) $\frac{19}{84}$

18) $\frac{52}{57}$

19) $4\frac{4}{27}$

20) $2\frac{26}{33}$

21) $3\frac{17}{36}$

22) $\frac{19}{22}$

23) $3\frac{27}{91}$

24) $3\frac{13}{36}$

25) $7\frac{1}{5}$

26) $4\frac{16}{45}$

WWW.MathNotion.Com

Chapter 3 :
Decimals

Topics that you will practice in this chapter:

- ✓ Adding and Subtracting Decimals
- ✓ Multiplying and Dividing Decimals
- ✓ Comparing Decimals
- ✓ Rounding Decimals

"The study of mathematics, like the Nile, begins in minuteness but ends in magnificence." –
Charles Caleb Colton

Adding and Subtracting Decimals

✎ Add and subtract decimals.

1) 35.19
   -24.28
   _____

2) 34.29
   $+42.58$
   _____

3) 61.20
   $+33.75$
   _____

4) 38.72
   -21.68
   _____

5) 57.39
   $+26.54$
   _____

6) 70.24
   -42.35
   _____

7) 86.09
   -35.14
   _____

8) 54.51
   $+32.66$
   _____

9) 114.21
   -88.69
   _____

✎ Find the missing number.

10) ___ + 2.8 = 5.4

11) 4.1 + ___ = 5.88

12) 6.45 + ___ = 8

13) 7.25 − ___ = 3.40

14) ___ − 2.35 = 4.25

15) ___ − 19.85 = 6.54

16) 22.15 + ___ = 28.95

17) ___ − 37.16 = 9.42

18) ___ + 24.50 = 34.19

19) 72.40 + ___ = 125.20

Multiplying and Dividing Decimals

✎ **Find the product.**

1) $0.5 \times 0.6 =$

2) $3.3 \times 0.4 =$

3) $1.28 \times 0.5 =$

4) $0.35 \times 0.6 =$

5) $1.85 \times 0.6 =$

6) $0.24 \times 0.5 =$

7) $5.25 \times 1.4 =$

8) $18.5 \times 4.6 =$

9) $15.4 \times 6.8 =$

10) $19.5 \times 2.6 =$

11) $32.2 \times 1.5 =$

12) $78.4 \times 4.5 =$

✎ **Find the quotient.**

13) $1.85 \div 10 =$

14) $74.6 \div 100 =$

15) $3.6 \div 3 =$

16) $9.6 \div 0.4 =$

17) $15.5 \div 0.5 =$

18) $32.8 \div 0.2 =$

19) $22.15 \div 1,000 =$

20) $53.55 \div 0.7 =$

21) $322.2 \div 0.2 =$

22) $50.67 \div 0.18 =$

23) $77.4 \div 0.8 =$

24) $27.93 \div 0.03 =$

Comparing Decimals

✏️ Write the correct comparison symbol (>, < or =).

1) 0.70 ☐ 0.070

2) 0.049 ☐ 0.49

3) 5.090 ☐ 5.09

4) 2.57 ☐ 2.05

5) 9.03 ☐ 0.930

6) 6.06 ☐ 6.6

7) 7.02 ☐ 7.020

8) 3.04 ☐ 3.2

9) 3.61 ☐ 3.245

10) 0.986 ☐ 0.0986

11) 17.24 ☐ 17.240

12) 0.759 ☐ 0.81

13) 9.040 ☐ 9.40

14) 5.73 ☐ 5.213

15) 9.44 ☐ 9.404

16) 7.17 ☐ 7.170

17) 4.85 ☐ 4.085

18) 9.041 ☐ 9.40

19) 3.033 ☐ 3.030

20) 4.97 ☐ 4.970

PSSA Subject Test Mathematics Grade 7

Rounding Decimals

✎ **Round each decimal to the nearest whole number.**

1) 28.12   3) 16.22   5) 7.95

2) 6.9   4) 8.5   6) 52.7

✎ **Round each decimal to the nearest tenth.**

7) 31.761   9) 94.729   11) 13.219

8) 14.421   10) 77.89   12) 59.89

✎ **Round each decimal to the nearest hundredth.**

13) 8.428   15) 55.3786   17) 62.241

14) 23.812   16) 231.912   18) 19.447

✎ **Round each decimal to the nearest thousandth.**

19) 15.54324   21) 243.8652   23) 67.1983

20) 34.62586   22) 80.4529   24) 72.36788

PSSA Subject Test Mathematics Grade 7

Answers of Worksheets

Adding and Subtracting Decimals

1) 10.91	6) 27.89	11) 1.78	16) 6.8
2) 76.87	7) 50.95	12) 1.55	17) 46.58
3) 94.95	8) 87.17	13) 3.85	18) 9.69
4) 17.04	9) 25.52	14) 6.6	19) 52.8
5) 83.93	10) 2.6	15) 26.39	

Multiplying and Dividing Decimals

1) 0.3	7) 7.35	13) 0.185	19) 0.02215
2) 1.32	8) 85.1	14) 0.746	20) 76.5
3) 0.64	9) 104.72	15) 1.2	21) 1,611
4) 0.21	10) 50.7	16) 24	22) 281.5
5) 1.11	11) 48.3	17) 31	23) 96.75
6) 0.12	12) 352.8	18) 164	24) 931

Comparing Decimals

1) >	6) <	11) =	16) =
2) <	7) =	12) <	17) >
3) =	8) <	13) <	18) <
4) >	9) >	14) >	19) >
5) >	10) >	15) >	20) =

Rounding Decimals

1) 28	9) 94.7	17) 62.24
2) 7	10) 77.9	18) 19.45
3) 16	11) 13.2	19) 15.543
4) 9	12) 59.9	20) 34.626
5) 8	13) 8.43	21) 243.865
6) 53	14) 23.81	22) 80.453
7) 31.8	15) 55.38	23) 67.198
8) 14.4	16) 231.91	24) 72.368

PSSA Subject Test Mathematics Grade 7

Chapter 4:
Proportions, Ratios, and Percent

Topics that you will practice in this chapter:

- ✓ Simplifying Ratios
- ✓ Proportional Ratios
- ✓ Similarity and Ratios
- ✓ Ratio and Rates Word Problems
- ✓ Percentage Calculations
- ✓ Percent Problems
- ✓ Discount, Tax and Tip
- ✓ Percent of Change
- ✓ Simple Interest

Without mathematics, there's nothing you can do. Everything around you is mathematics. Everything around you is numbers." – Shakuntala Devi

Simplifying Ratios

✍ Reduce each ratio.

1) $15:20 = \_\_:\_\_$
2) $7:70 = \_\_:\_\_$
3) $16:28 = \_\_:\_\_$
4) $7:21 = \_\_:\_\_$
5) $4:40 = \_\_:\_\_$
6) $6:48 = \_\_:\_\_$
7) $16:64 = \_\_:\_\_$
8) $10:25 = \_\_:\_\_$

9) $8:48 = \_\_:\_\_$
10) $49:63 = \_\_:\_\_$
11) $18:27 = \_\_:\_\_$
12) $35:10 = \_\_:\_\_$
13) $90:9 = \_\_:\_\_$
14) $24:32 = \_\_:\_\_$
15) $7:56 = \_\_:\_\_$
16) $45:63 = \_\_:\_\_$

17) $56:72 = \_\_:\_\_$
18) $26:13 = \_\_:\_\_$
19) $15:45 = \_\_:\_\_$
20) $28:4 = \_\_:\_\_$
21) $24:48 = \_\_:\_\_$
22) $30:24 = \_\_:\_\_$
23) $70:140 = \_\_:\_\_$
24) $6:180 = \_\_:\_\_$

✍ Write each ratio as a fraction in simplest form.

25) $6:12 =$
26) $30:50 =$
27) $15:35 =$
28) $9:27 =$
29) $8:24 =$
30) $18:84 =$
31) $7:14 =$

32) $7:35 =$
33) $40:96 =$
34) $12:54 =$
35) $44:52 =$
36) $12:27 =$
37) $15:180 =$
38) $39:143 =$

39) $20:300 =$
40) $30:120 =$
41) $56:42 =$
42) $26:130 =$
43) $66:123 =$
44) $70:630 =$
45) $75:125 =$

PSSA Subject Test Mathematics Grade 7

Proportional Ratios

✎ Fill in the blanks; Calculate each proportion.

1) $3:8 = \underline{\quad}:48$

2) $2:5 = 20:\underline{\quad}$

3) $1:9 = \underline{\quad}:81$

4) $6:7 = 12:\underline{\quad}$

5) $9:2 = 63:\underline{\quad}$

6) $8:7 = \underline{\quad}:49$

7) $20:3 = \underline{\quad}:15$

8) $1:3 = \underline{\quad}:75$

9) $7:6 = \underline{\quad}:60$

10) $8:5 = \underline{\quad}:45$

11) $3:10 = 60:\underline{\quad}$

12) $6:11 = 42:\underline{\quad}$

✎ State if each pair of ratios form a proportion.

13) $\frac{3}{20}$ and $\frac{9}{60}$

14) $\frac{1}{7}$ and $\frac{6}{42}$

15) $\frac{3}{7}$ and $\frac{24}{56}$

16) $\frac{4}{9}$ and $\frac{12}{18}$

17) $\frac{1}{9}$ and $\frac{12}{81}$

18) $\frac{7}{8}$ and $\frac{21}{28}$

19) $\frac{9}{13}$ and $\frac{27}{39}$

20) $\frac{1}{8}$ and $\frac{8}{64}$

21) $\frac{6}{19}$ and $\frac{30}{85}$

22) $\frac{5}{9}$ and $\frac{40}{81}$

23) $\frac{9}{14}$ and $\frac{108}{168}$

24) $\frac{15}{23}$ and $\frac{360}{552}$

✎ Calculate each proportion.

25) $\frac{20}{25} = \frac{32}{x}, x = \underline{\quad}$

26) $\frac{1}{8} = \frac{32}{x}, x = \underline{\quad}$

27) $\frac{15}{5} = \frac{21}{x}, x = \underline{\quad}$

28) $\frac{1}{7} = \frac{x}{294}, x = \underline{\quad}$

29) $\frac{7}{9} = \frac{x}{81}, x = \underline{\quad}$

30) $\frac{1}{5} = \frac{13}{x}, x = \underline{\quad}$

31) $\frac{9}{5} = \frac{36}{x}, x = \underline{\quad}$

32) $\frac{6}{13} = \frac{48}{x}, x = \underline{\quad}$

33) $\frac{5}{8} = \frac{x}{88}, x = \underline{\quad}$

34) $\frac{4}{15} = \frac{x}{240}, x = \underline{\quad}$

35) $\frac{9}{19} = \frac{x}{266}, x = \underline{\quad}$

36) $\frac{7}{15} = \frac{x}{270}, x = \underline{\quad}$

WWW.MathNotion.Com

PSSA Subject Test Mathematics Grade 7

Similarity and Ratios

✍ **Each pair of figures is similar. Find the missing side.**

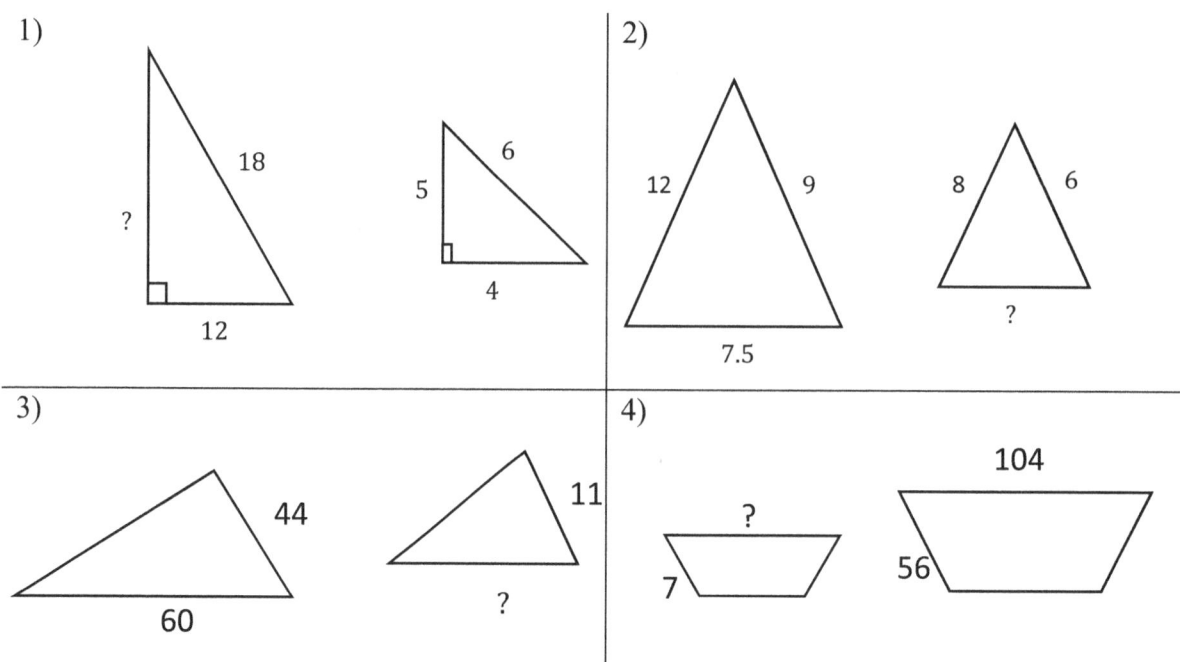

✍ **Calculate.**

5) Two rectangles are similar. The first is 24 feet wide and 120 feet long. The second is 30 feet wide. What is the length of the second rectangle? _____

6) Two rectangles are similar. One is 5 meters by 36 meters. The longer side of the second rectangle is 90 meters. What is the other side of the second rectangle? _____

7) A building casts a shadow 25 ft long. At the same time a girl 10 ft tall casts a shadow 5 ft long. How tall is the building? _____

8) The scale of a map of Texas is 4 inches: 32 miles. If you measure the distance from Dallas to Martin County as 38.4 inches, approximately how far is Martin County from Dallas? _____

PSSA Subject Test Mathematics Grade 7

Ratio and Rates Word Problems

✎ Find the answer for each word problem.

1) Mason has 24 red cards and 36 green cards. What is the ratio of Mason's red cards to his green cards? _____

2) In a party, 45 soft drinks are required for every 54 guests. If there are 378 guests, how many soft drinks is required? _____

3) In Mason's class, 42 of the students are tall and 24 are short. In Michael's class 84 students are tall and 48 students are short. Which class has a higher ratio of tall to short students? _____

4) The price of 5 apples at the Quick Market is $4.6. The price of 7 of the same apples at Walmart is $5.95. Which place is the better buy? _____

5) The bakers at a Bakery can make 90 bagels in 3 hours. How many bagels can they bake in 24 hours? What is that rate per hour? _____

6) You can buy 5 cans of green beans at a supermarket for $5.75. How much does it cost to buy 45 cans of green beans? _____

7) The ratio of boys to girls in a class is 4: 7. If there are 32 boys in the class, how many girls are in that class? _____

8) The ratio of red marbles to blue marbles in a bag is 3: 7. If there are 50 marbles in the bag, how many of the marbles are red? _____

WWW.MathNotion.Com

PSSA Subject Test Mathematics Grade 7

Percentage Calculations

✍ Calculate the given percent of each value.

1) 3% of 60 = ___
2) 20% of 32 = ___
3) 4% of 72 = ___
4) 16% of 32 = ___
5) 25% of 124 = ___
6) 35% of 56 = ___

7) 15% of 20 = ___
8) 14% of 150 = ___
9) 80% of 50 = ___
10) 12% of 115 = ___
11) 72% of 250 = ___
12) 52% of 500 = ___

13) 70% of 400 = ___
14) 27% of 145 = ___
15) 90% of 64 = ___
16) 60% of 55 = ___
17) 22% of 210 = ___
18) 8% of 235 = ___

✍ Calculate the percent of each given value.

19) ___% of 25 = 5
20) ___% of 40 = 20
21) ___% of 25 = 2
22) ___% of 50 = 16
23) ___% of 250 = 5

24) ___% of 40 = 32
25) ___% of 125 = 20
26) ___% of 700 = 49
27) ___% of 350 = 49
28) ___% of 500 = 210

✍ Calculate each percent problem.

29) A Cinema has 250 seats. 60 seats were sold for the current movie. What percent of seats are empty? ___ %

30) There are 68 boys and 92 girls in a class. 75% of the students in the class take the bus to school. How many students do not take the bus to school? ___

PSSA Subject Test Mathematics Grade 7

Percent Problems

✎ **Calculate each problem.**

1) 9 is what percent of 45? ___%

2) 60 is what percent of 120? ___%

3) 10 is what percent of 200? ___%

4) 15 is what percent of 125? ___%

5) 10 is what percent of 400? ___%

6) 66 is what percent of 55? ___%

7) 40 is what percent of 160? ___%

8) 40 is what percent of 50? ___%

9) 120 is what percent of 800? ___%

10) 78 is what percent of 120? ___%

11) 36 is what percent of 144? ___%

12) 17 is what percent of 85? ___%

13) 90 is what percent of 900? ___%

14) 36 is what percent of 16? ___%

15) 63 is what percent of 14? ___%

16) 18 is what percent of 60? ___%

17) 126 is what percent of 200? ___%

18) 232 is what percent of 40? ___%

✎ **Calculate each percent word problem.**

19) There are 40 employees in a company. On a certain day, 25 were present. What percent showed up for work? ____%

20) A metal bar weighs 60 ounces. 25% of the bar is gold. How many ounces of gold are in the bar? _____

21) A crew is made up of 12 women; the rest are men. If 15% of the crew are women, how many people are in the crew? _____

22) There are 40 students in a class and 8 of them are girls. What percent are boys? ____%

23) The Royals softball team played 400 games and won 280 of them. What percent of the games did they lose? ____%

PSSA Subject Test Mathematics Grade 7

Discount, Tax and Tip

✍ **Find the selling price of each item.**

1) Original price of a computer: $420
   Tax: 8% Selling price: $_____

2) Original price of a laptop: $280
   Tax: 4% Selling price: $_____

3) Original price of a sofa: $820
   Tax: 5% Selling price: $_____

4) Original price of a car: $15,800
   Tax: 3.6%   Selling price: $_____

5) Original price of a Table: $250
   Tax: 9% Selling price: $_____

6) Original price of a house: $630,000
   Tax: 1.8%   Selling price: $_____

7) Original price of a tablet: $450
   Discount: 30%   Selling price: $____

8) Original price of a chair: $390
   Discount: 8%   Selling price: $____

9) Original price of a book: $75
   Discount: 42%   Selling price: $____

10) Original price of a cellphone: $820
    Discount: 23%   Selling price: $___

11) Food bill: $45
    Tip: 15%   Price: $_____

12) Food bill: $32
    Tipp: 20%   Price: $_____

13) Food bill: $90
    Tip: 35%   Price: $_____

14) Food bill: $42
    Tipp: 12%   Price: $_____

✍ **Find the answer for each word problem.**

15) Nicolas hired a moving company. The company charged $500 for its services, and Nicolas gives the movers a 40% tip. How much does Nicolas tip the movers? $_____

16) Mason has lunch at a restaurant and the cost of his meal is $90. Mason wants to leave a 25% tip. What is Mason's total bill including tip? $_____

17) The sales tax in a state is 19.80% and an item costs $350. How much is the tax? $_____

18) The price of a table at Best Buy is $680. If the sales tax is 5%, what is the final price of the table including tax? $_____

WWW.MathNotion.Com

PSSA Subject Test Mathematics Grade 7

Percent of Change

✎ **Find each percent of change.**

1) From 150 to 450. ___ %

2) From 50 ft to 250 ft. ___ %

3) From $60 to $360. ___ %

4) From 60 cm to 180 cm. ___ %

5) From 15 to 45. ___ %

6) From 80 to 16. ___ %

7) From 120 to 360. ___ %

8) From 900 to 450. ___ %

9) From 1,000 to 200. ___ %

10) From 144 to 36. ___ %

✎ **Calculate each percent of change word problem.**

11) Bob got a raise, and his hourly wage increased from $42 to $63. What is the percent increase? ___ %

12) The price of a pair of shoes increases from $50 to $61. What is the percent increase? ___ %

13) At a coffee shop, the price of a cup of coffee increased from $4.80 to $5.76. What is the percent increase in the cost of the coffee? ___ %

14) 51 cm are cut from 85 cm board. What is the percent decrease in length? ___ %

15) In a class, the number of students has been increased from 54 to 81. What is the percent increase? ___ %

16) The price of gasoline rises from $24.40 to $30.50 in one month. By what percent did the gas price rise? ___ %

17) A shirt was originally priced at $38. It went on sale for $24.70. What was the percent that the shirt was discounted? ___ %

WWW.MathNotion.Com

PSSA Subject Test Mathematics Grade 7

Simple Interest

✍ **Determine the simple interest for these loans.**

1) $480 at 11% for 3 years. $ _____

2) $4,200 at 7% for 4 years. $ _____

3) $2,500 at 20% for 3 years. $ _____

4) $6,800 at 3.9% for 4 months. $ _____

5) $800 at 6% for 7 months. $ _____

6) $36,000 at 4.2% for 6 years. $ _____

7) $6,500 at 7% for 4 years. $ _____

8) $850 at 9.5% for 2 years. $ _____

9) $1,200 at 5.8% for 9 months. $ _____

10) $3,000 at 4.5% for 7 years. $ _____

✍ **Calculate each simple interest word problem.**

11) A new car, valued at $22,000, depreciates at 8.5% per year. What is the value of the car one year after purchase? $_____

12) Sara puts $9,000 into an investment yielding 6% annual simple interest; she left the money in for three years. How much interest does Sara get at the end of those three years? $_____

13) A bank is offering 12% simple interest on a savings account. If you deposit $16,400, how much interest will you earn in two years? $_____

14) $720 interest is earned on a principal of $6,000 at a simple interest rate of 4% interest per year. For how many years was the principal invested? _____

15) In how many years will $2,200 yield an interest of $440 at 4% simple interest? _____

16) Jim invested $8,000 in a bond at a yearly rate of 4.5%. He earned $1,440 in interest. How long was the money invested? _____

PSSA Subject Test Mathematics Grade 7

Answers of Worksheets

Simplifying Ratios

1) 3:4
2) 1:10
3) 4:7
4) 1:3
5) 1:10
6) 1:8
7) 2:8
8) 2:5
9) 1:6
10) 7:9
11) 2:3
12) 7:2
13) 10:1
14) 3:4
15) 1:8
16) 5:7
17) 7:9
18) 2:1
19) 1:3
20) 7:1
21) 1:2
22) 5:4
23) 1:2
24) 1:30
25) $\frac{1}{2}$
26) $\frac{3}{5}$
27) $\frac{3}{7}$
28) $\frac{1}{3}$
29) $\frac{1}{3}$
30) $\frac{3}{14}$
31) $\frac{1}{2}$
32) $\frac{1}{5}$
33) $\frac{5}{12}$
34) $\frac{2}{9}$
35) $\frac{11}{13}$
36) $\frac{4}{9}$
37) $\frac{1}{12}$
38) $\frac{3}{11}$
39) $\frac{1}{15}$
40) $\frac{1}{4}$
41) $\frac{4}{3}$
42) $\frac{1}{5}$
43) $\frac{22}{41}$
44) $\frac{1}{9}$
45) $\frac{3}{5}$

Proportional Ratios

1) 18
2) 50
3) 9
4) 14
5) 14
6) 56
7) 100
8) 25
9) 70
10) 72
11) 200
12) 77
13) Yes
14) Yes
15) Yes
16) No
17) No
18) No
19) Yes
20) Yes
21) No
22) No
23) Yes
24) Yes
25) 40
26) 256
27) 7
28) 42
29) 63
30) 65
31) 20
32) 104
33) 55
34) 64
35) 126
36) 126

Similarity and ratios

1) 15
2) 5
3) 15
4) 13
5) 150 feet
6) 12.5 meters
7) 50 feet
8) 307.2 miles

Ratio and Rates Word Problems

1) 2:3
2) 315

WWW.MathNotion.Com

PSSA Subject Test Mathematics Grade 7

3) The ratio for both classes is 7 to 4. 6) $51.75
4) Walmart is a better buy. 7) 56
5) 720, the rate is 30 per hour. 8) 15

Percentage Calculations

1) 1.8   11) 180   21) 8%
2) 6.4   12) 260   22) 32%
3) 2.88   13) 280   23) 2%
4) 5.12   14) 39.15   24) 80%
5) 31   15) 57.6   25) 16%
6) 19.6   16) 33   26) 7%
7) 3   17) 46.2   27) 14%
8) 21   18) 18.8   28) 42%
9) 40   19) 20%   29) 76%
10) 13.8   20) 50%   30) 40

Percent Problems

1) 20%   9) 15%   17) 63%
2) 50%   10) 65%   18) 580%
3) 5%   11) 25%   19) 62.5%
4) 12%   12) 20%   20) 15 ounces
5) 2.5%   13) 10%   21) 80
6) 120%   14) 225%   22) 80%
7) 25%   15) 450%   23) 30%
8) 80%   16) 30%

Discount, Tax and Tip

1) $453.60   7) $315.00   13) $121.50
2) $291.20   8) $358.80   14) $47.04
3) $861.00   9) $43.50   15) $200.00
4) $16,368.80   10) $631.40   16) $112.50
5) $272.50   11) $51.75   17) $69.30
6) $641,340   12) $38.40   18) $714.00

PSSA Subject Test Mathematics Grade 7

Percent of Change

1) 200%
2) 400%
3) 500%
4) 200%
5) 200%
6) 80%
7) 200%
8) 50%
9) 80%
10) 75%
11) 50%
12) 22%
13) 20%
14) 60%
15) 50%
16) 25%
17) 35%

Simple Interest

1) $158.40
2) $1,176.00
3) $1,500.00
4) $88.40
5) $28.00
6) $9,072.00
7) $1,820.00
8) $161.50
9) $52.20
10) $945.00
11) $20,130.00
12) $1,620.00
13) $3,936.00
14) 3 years
15) 5 years
16) 4 years

Chapter 5:
Exponents and Radicals Expressions

Topics that you will practice in this chapter:

- ✓ Multiplication Property of Exponents
- ✓ Zero and Negative Exponents
- ✓ Division Property of Exponents
- ✓ Powers of Products and Quotients
- ✓ Negative Exponents and Negative Bases
- ✓ Scientific Notation
- ✓ Square Roots

Mathematics is no more computation than typing is literature.

– John Allen Paulos

Multiplication Property of Exponents

✎ Simplify and write the answer in exponential form.

1) $4 \times 4^5 =$

2) $8^4 \times 8 =$

3) $7^3 \times 7^3 =$

4) $9^2 \times 9^2 =$

5) $2^2 \times 2^4 \times 2 =$

6) $5 \times 5^3 \times 5^3 =$

7) $4^3 \times 4^2 \times 4 \times 4 =$

8) $5x \times x =$

9) $x^3 \times x^3 =$

10) $x^7 \times x^2 =$

11) $x^4 \times x^3 \times x^2 =$

12) $10x \times 3x =$

13) $4x^3 \times 4x^3 =$

14) $7x^3 \times x =$

15) $3x^2 \times 4x^2 \times x^2 =$

16) $5x^4 \times x^4 =$

17) $2x^8 \times 2x =$

18) $6x \times x^5 =$

19) $4x^2 \times 6x^6 =$

20) $5yx^3 \times 4x =$

21) $7x^3 \times y^5 x^7 =$

22) $y^2 x^3 \times y^5 x^4 =$

23) $3x^5 \times 4x^3 y^4 =$

24) $4x^4 \times 9x^2 y^5 =$

25) $5x^3 y^4 \times 6x^8 y^2 =$

26) $8x^3 y^6 \times 4xy^3 =$

27) $2xy^5 \times 6x^3 y^3 =$

28) $4x^5 y^2 \times 4x^2 y^8 =$

29) $7x \times 3y^8 x^2 \times y^5 =$

30) $x^3 \times 2y^3 x^4 \times 2y =$

31) $3yx^4 \times 3y^4 x \times 3xy^3 =$

32) $6y^3 \times 2y^2 x^4 \times 10yx^5 =$

PSSA Subject Test Mathematics Grade 7

Zero and Negative Exponents

✎ **Evaluate the following expressions.**

1) $1^{-5} =$

2) $4^{-1} =$

3) $0^{10} =$

4) $1^{15} =$

5) $5^{-2} =$

6) $3^{-3} =$

7) $9^{-1} =$

8) $10^{-2} =$

9) $12^{-2} =$

10) $2^{-5} =$

11) $3^{-4} =$

12) $2^{-4} =$

13) $6^{-3} =$

14) $10^{-3} =$

15) $30^{-1} =$

16) $15^{-2} =$

17) $4^{-3} =$

18) $2^{-7} =$

19) $5^{-3} =$

20) $4^{-4} =$

21) $3^{-5} =$

22) $10^{-4} =$

23) $2^{-10} =$

24) $8^{-3} =$

25) $20^{-2} =$

26) $14^{-2} =$

27) $9^{-3} =$

28) $100^{-2} =$

29) $5^{-4} =$

30) $4^{-6} =$

31) $\left(\frac{1}{4}\right)^{-3} =$

32) $\left(\frac{1}{6}\right)^{-2} =$

33) $\left(\frac{1}{7}\right)^{-2} =$

34) $\left(\frac{2}{3}\right)^{-3} =$

35) $\left(\frac{1}{13}\right)^{-2} =$

36) $\left(\frac{7}{12}\right)^{-2} =$

37) $\left(\frac{1}{6}\right)^{-3} =$

38) $\left(\frac{1}{300}\right)^{-2} =$

39) $\left(\frac{2}{9}\right)^{-2} =$

40) $\left(\frac{7}{5}\right)^{-1} =$

41) $\left(\frac{13}{23}\right)^{0} =$

42) $\left(\frac{1}{4}\right)^{-5} =$

WWW.MathNotion.Com

PSSA Subject Test Mathematics Grade 7

Division Property of Exponents

✏️ **Simplify.**

1) $\dfrac{5^6}{5^7} =$

2) $\dfrac{8^8}{8^6} =$

3) $\dfrac{4^5}{4} =$

4) $\dfrac{3}{3^5} =$

5) $\dfrac{x}{x^6} =$

6) $\dfrac{3 \times 3^2}{3^2 \times 3^5} =$

7) $\dfrac{9^4}{9^2} =$

8) $\dfrac{10 \times 10^9}{10^2 \times 10^7} =$

9) $\dfrac{7^5 \times 7^7}{7^4 \times 7^8} =$

10) $\dfrac{15x}{30x^6} =$

11) $\dfrac{3x^9}{4x^4} =$

12) $\dfrac{15x^8}{10x^9} =$

13) $\dfrac{42x^5}{6y^9} =$

14) $\dfrac{36y^8}{4x^4y^5} =$

15) $\dfrac{2x^7}{9x} =$

16) $\dfrac{49x^8y^6}{7x^9} =$

17) $\dfrac{48x^2}{24x^6y^{12}} =$

18) $\dfrac{30yx^5}{6yx^7} =$

19) $\dfrac{19x^7y}{38x^{12}y^4} =$

20) $\dfrac{9x^8}{63x^8} =$

21) $\dfrac{9x^{-9}}{4x^{-3}} =$

Powers of Products and Quotients

✏️ **Simplify.**

1) $(4^3)^2 =$

2) $(2^3)^4 =$

3) $(2 \times 2^3)^2 =$

4) $(5 \times 5^5)^6 =$

5) $(19^4 \times 19^2)^3 =$

6) $(2^3 \times 2^4)^4 =$

7) $(5 \times 5^2)^2 =$

8) $(4^4)^4 =$

9) $(8x^5)^2 =$

10) $(3x^2y^4)^4 =$

11) $(7x^5y^2)^2 =$

12) $(5x^4y^4)^3 =$

13) $(2x^3y^3)^5 =$

14) $(10x^3y^4)^3 =$

15) $(13y^3y)^2 =$

16) $(5x^6x^4)^2 =$

17) $(6x^7y^6)^3 =$

18) $(12x^5x^7)^2 =$

19) $(2x^4 \times 2x)^4 =$

20) $(2x^4y^3)^5 =$

21) $(15x^7y^2)^2 =$

22) $(8x^3y^5)^3 =$

23) $(3x \times 2y^2)^4 =$

24) $\left(\dfrac{4x}{x^5}\right)^2 =$

25) $\left(\dfrac{x^4y^5}{x^3y^5}\right)^9 =$

26) $\left(\dfrac{36xy}{6x^5}\right)^3 =$

27) $\left(\dfrac{x^7}{x^8y^2}\right)^6 =$

28) $\left(\dfrac{xy^4}{x^3y^6}\right)^{-3} =$

29) $\left(\dfrac{5xy^8}{x^3}\right)^2 =$

30) $\left(\dfrac{xy^6}{2xy^3}\right)^{-4} =$

PSSA Subject Test Mathematics Grade 7

Negative Exponents and Negative Bases

✏️ **Simplify.**

1) $-9^{-1} =$

2) $-9^{-2} =$

3) $-2^{-5} =$

4) $-x^{-7} =$

5) $11x^{-1} =$

6) $-8x^{-3} =$

7) $-12x^{-5} =$

8) $-9x^{-8}y^{-6} =$

9) $32x^{-5}y^{-1} =$

10) $10a^{-9}b^{-3} =$

11) $-17x^4y^{-6} =$

12) $-\dfrac{25}{x^{-5}} =$

13) $-\dfrac{13x}{a^{-7}} =$

14) $\left(-\dfrac{1}{3}\right)^{-4} =$

15) $\left(-\dfrac{3}{4}\right)^{-2} =$

16) $-\dfrac{14}{a^{-6}b^{-3}} =$

17) $-\dfrac{7x}{x^{-8}} =$

18) $-\dfrac{a^{-9}}{b^{-5}} =$

19) $-\dfrac{11}{x^{-5}} =$

20) $\dfrac{8b}{-16c^{-6}} =$

21) $\dfrac{12ab}{a^{-4}b^{-3}} =$

22) $-\dfrac{8n^{-4}}{32p^{-7}} =$

23) $\dfrac{16ab^{-6}}{-6c^{-5}} =$

24) $\left(\dfrac{10a}{5c}\right)^{-4} =$

25) $\left(-\dfrac{12x}{4yz}\right)^{-3} =$

26) $\dfrac{8ab^{-7}}{-5c^{-3}} =$

27) $\left(-\dfrac{x^4}{x^5}\right)^{-5} =$

28) $\left(-\dfrac{x^{-2}}{7x^3}\right)^{-2} =$

29) $\left(-\dfrac{x^{-4}}{x^2}\right)^{-6} =$

Scientific Notation

✎ Write each number in scientific notation.

1) $0.223 =$

2) $0.09 =$

3) $4.5 =$

4) $900 =$

5) $2,000 =$

6) $0.006 =$

7) $33 =$

8) $9,400 =$

9) $1,470 =$

10) $52,000 =$

11) $8,000,000 =$

12) $0.00009 =$

13) $2,158,000 =$

14) $0.0039 =$

15) $0.000075 =$

16) $4,300,000 =$

17) $130,000 =$

18) $4,000,000,000 =$

19) $0.00009 =$

20) $0.0039 =$

✎ Write each number in standard notation.

21) $4 \times 10^{-1} =$

22) $1.2 \times 10^{-3} =$

23) $2.7 \times 10^{5} =$

24) $6 \times 10^{-4} =$

25) $3.6 \times 10^{-3} =$

26) $5.5 \times 10^{5} =$

27) $3.2 \times 10^{4} =$

28) $3.88 \times 10^{6} =$

29) $7 \times 10^{-6} =$

30) $4.2 \times 10^{-7} =$

PSSA Subject Test Mathematics Grade 7

Square Roots

✎ Find the value each square root.

1) $\sqrt{16} =$ ___

2) $\sqrt{25} =$ ___

3) $\sqrt{1} =$ ___

4) $\sqrt{64} =$ ___

5) $\sqrt{0} =$ ___

6) $\sqrt{196} =$ ___

7) $\sqrt{4} =$ ___

8) $\sqrt{256} =$ ___

9) $\sqrt{36} =$ ___

10) $\sqrt{289} =$ ___

11) $\sqrt{169} =$ ___

12) $\sqrt{144} =$ ___

13) $\sqrt{100} =$ ___

14) $\sqrt{1,600} =$ ___

15) $\sqrt{2,500} =$ ___

16) $\sqrt{324} =$ ___

17) $\sqrt{529} =$ ___

18) $\sqrt{20} =$ ___

19) $\sqrt{625} =$ ___

20) $\sqrt{18} =$ ___

21) $\sqrt{50} =$ ___

22) $\sqrt{1,024} =$ ___

23) $\sqrt{160} =$ ___

24) $\sqrt{32} =$ ___

✎ Evaluate.

25) $\sqrt{4} \times \sqrt{25} =$ _____

26) $\sqrt{36} \times \sqrt{49} =$ _____

27) $\sqrt{6} \times \sqrt{6} =$ _____

28) $\sqrt{13} \times \sqrt{13} =$ _____

29) $2\sqrt{5} \times 3\sqrt{5} =$ _____

30) $\sqrt{12} \times \sqrt{3} =$ _____

31) $\sqrt{13} + \sqrt{13} =$ _____

32) $\sqrt{10} + 2\sqrt{10} =$ _____

33) $12\sqrt{7} - 10\sqrt{7} =$ _____

34) $4\sqrt{10} \times 2\sqrt{10} =$ _____

35) $5\sqrt{3} \times 8\sqrt{3} =$ _____

36) $6\sqrt{3} - \sqrt{12} =$ _____

PSSA Subject Test Mathematics Grade 7

Answers of Worksheets

Multiplication Property of Exponents

1) 4^6
2) 8^5
3) 7^6
4) 9^4
5) 2^7
6) 5^7
7) 4^7
8) $5x^2$
9) x^6
10) x^9
11) x^9
12) $30x^2$
13) $16x^6$
14) $7x^4$
15) $12x^6$
16) $5x^8$
17) $4x^9$
18) $6x^6$
19) $24x^8$
20) $20x^4y$
21) $7x^{10}y^5$
22) x^7y^7
23) $12x^8y^4$
24) $36x^6y^5$
25) $30x^{11}y^6$
26) $32x^4y^9$
27) $12x^4y^8$
28) $16x^7y^{10}$
29) $21x^3y^{13}$
30) $4x^7y^4$
31) $27x^6y^8$
32) $120x^9y^6$

Zero and Negative Exponents

1) 1
2) $\frac{1}{4}$
3) 0
4) 1
5) $\frac{1}{25}$
6) $\frac{1}{27}$
7) $\frac{1}{9}$
8) $\frac{1}{100}$
9) $\frac{1}{144}$
10) $\frac{1}{32}$
11) $\frac{1}{81}$
12) $\frac{1}{16}$
13) $\frac{1}{216}$
14) $\frac{1}{1,000}$
15) $\frac{1}{30}$
16) $\frac{1}{225}$
17) $\frac{1}{64}$
18) $\frac{1}{128}$
19) $\frac{1}{125}$
20) $\frac{1}{256}$
21) $\frac{1}{243}$
22) $\frac{1}{10,000}$
23) $\frac{1}{1,024}$
24) $\frac{1}{512}$
25) $\frac{1}{400}$
26) $\frac{1}{196}$
27) $\frac{1}{729}$
28) $\frac{1}{10,000}$
29) $\frac{1}{625}$
30) $\frac{1}{4,096}$
31) 64
32) 36
33) 49
34) $\frac{27}{8}$
35) 169
36) $\frac{144}{49}$
37) 216
38) $90,000$
39) $\frac{81}{4}$
40) $\frac{5}{7}$
41) 1
42) $1,024$

Division Property of Exponents

1) $\frac{1}{5}$
2) 8^2
3) 4^4
4) $\frac{1}{3^4}$
5) $\frac{1}{x^5}$
6) $\frac{1}{3^4}$
7) 9^2
8) 10
9) 1
10) $\frac{1}{2x^5}$
11) $\frac{3x^5}{4}$
12) $\frac{3}{2x}$
13) $\frac{7x^5}{y^9}$
14) $\frac{9y^3}{x^4}$
15) $\frac{2x^6}{9}$

PSSA Subject Test Mathematics Grade 7

16) $\frac{7y^6}{x}$
17) $\frac{2}{x^4 y^{12}}$
18) $\frac{5}{x^2}$
19) $\frac{1}{2x^5 y^3}$
20) $\frac{1}{7}$
21) $\frac{9}{4x^6}$

Powers of Products and Quotients

1) 4^6
2) 2^{12}
3) 2^8
4) 5^{36}
5) 19^{18}
6) 2^{28}
7) 5^6
8) 4^{16}
9) $64x^{10}$
10) $81x^8 y^{16}$
11) $49x^{10} y^4$
12) $125x^{12} y^{12}$
13) $32x^{15} y^{15}$
14) $1{,}000 x^9 y^{12}$
15) $169 y^8$
16) $25 x^{20}$
17) $216 x^{21} y^{18}$
18) $144 x^{24}$
19) $256 x^{20}$
20) $32 x^{20} y^{15}$
21) $225 x^{14} y^4$
22) $512 x^9 y^{15}$
23) $1{,}296 x^4 y^8$
24) $\frac{16}{x^8}$
25) x^9
26) $\frac{216 y^3}{x^{12}}$
27) $\frac{1}{x^6 y^{12}}$
28) $x^6 y^6$
29) $\frac{25 y^{16}}{x^4}$
30) $\frac{16}{y^{12}}$

Negative Exponents and Negative Bases

1) $-\frac{1}{9}$
2) $-\frac{1}{81}$
3) $-\frac{1}{32}$
4) $-\frac{1}{x^7}$
5) $\frac{11}{x}$
6) $-\frac{8}{x^3}$
7) $-\frac{12}{x^5}$
8) $-\frac{9}{x^8 y^6}$
9) $\frac{32}{x^5 y}$
10) $\frac{10}{a^9 b^3}$
11) $-\frac{17 x^4}{y^6}$
12) $-25 x^5$
13) $-13 x a^7$
14) 81
15) $\frac{16}{9}$
16) $-14 a^6 b^3$
17) $-7 x^9$
18) $-\frac{b^5}{a^9}$
19) $-11 x^5$
20) $-\frac{bc^6}{2}$
21) $12 a^5 b^4$
22) $-\frac{p^7}{4 n^4}$
23) $-\frac{8 ac^5}{3 b^6}$
24) $\frac{c^4}{16 a^4}$
25) $\frac{y^3 z^3}{27 x^3}$
26) $-\frac{8 ac^3}{5 b^7}$
27) $-x^5$
28) $49 x^{10}$
29) x^{36}

Scientific Notation

PSSA Subject Test Mathematics Grade 7

1) 2.23×10^{-1}
2) 9×10^{-2}
3) 4.5×10^{0}
4) 9×10^{2}
5) 2×10^{3}
6) 6×10^{-3}
7) 3.3×10^{1}
8) 9.4×10^{3}
9) 1.47×10^{3}
10) 5.2×10^{4}
11) 8×10^{6}
12) 9×10^{-5}
13) 2.158×10^{6}
14) 3.9×10^{-3}
15) 7.5×10^{-5}
16) 4.3×10^{6}
17) 1.3×10^{5}
18) 4×10^{9}
19) 9×10^{-5}
20) 3.9×10^{-3}
21) 0.4
22) 0.0012
23) 270,000
24) 0.0006
25) 0.0036
26) 550,000
27) 32,000
28) 3,880,000
29) 0.000007
30) 0.00000042

Square Roots

1) 4
2) 5
3) 1
4) 8
5) 0
6) 14
7) 2
8) 16
9) 6
10) 17
11) 13
12) 12
13) 10
14) 40
15) 50
16) 18
17) 23
18) $2\sqrt{5}$
19) 25
20) $3\sqrt{2}$
21) $5\sqrt{2}$
22) 32
23) $4\sqrt{10}$
24) $4\sqrt{2}$
25) 10
26) 42
27) 6
28) 13
29) 30
30) 6
31) $2\sqrt{13}$
32) $3\sqrt{10}$
33) $2\sqrt{7}$
34) 80
35) 120
36) $4\sqrt{3}$

PSSA Subject Test Mathematics Grade 7

Chapter 6:
Algebraic Expressions

Topics that you will practice in this chapter:

- ✓ Translate Phrases into an Algebraic Statement
- ✓ Simplifying Variable Expressions
- ✓ The Distributive Property
- ✓ Evaluating One Variable Expressions
- ✓ Evaluating Two Variables Expressions
- ✓ Combining like Terms

Mathematics is, as it were, a sensuous logic, and relates to philosophy as do the arts, music, and plastic art to poetry. — K. Shegel

PSSA Subject Test Mathematics Grade 7

Translate Phrases into an Algebraic Statement

✏ Write an algebraic expression for each phrase.

1) 9 multiplied by x. _____

2) Subtract 11 from y. _____

3) 19 divided by x. _____

4) 38 decreased by y. _____

5) Add y to 40. _____

6) The square of 6. _____

7) x raised to the fifth power. _____

8) The sum of six and a number. _____

9) The difference between fifty-seven and y. _____

10) The quotient of nine and a number. _____

11) The quotient of the square of x and 25. _____

12) The difference between x and 6 is 19. _____

13) 10 times a reduced by the square of b. _____

14) Subtract the product of a and b from 41. _____

WWW.MathNotion.Com

Simplifying Variable Expressions

✏ Simplify each expression.

1) $3(x + 5) =$

2) $(-4)(7x - 5) =$

3) $11x + 5 - 6x =$

4) $-4 - 2x^2 - 6x^2 =$

5) $7 + 13x^2 + 3 =$

6) $3x^2 + 7x + 15x^2 =$

7) $3x^2 - 12x^2 + 4x =$

8) $4x^2 - 8x - 2x =$

9) $6x + 7(3 - 4x) =$

10) $8x + 4(15x - 3) =$

11) $6(-3x - 9) - 17 =$

12) $-11x^2 - (-5x) =$

13) $2x + 7 + 5 - 8x =$

14) $7 + 6x - 11 - 5x =$

15) $27x + 8 - 13 - 5x =$

16) $(-11)(-5x + 2) - 41x =$

17) $19x - 4(4 - 2x) =$

18) $16x + 3(3x + 6) + 10 =$

19) $5(-2x - 4) - 13x =$

20) $16x - 3x(x + 10) =$

21) $17x + 5x(2 - 4x) =$

22) $5x(-4x - 7) + 20x =$

23) $25x - 19 + 4x^2 =$

24) $6x(x - 11) + 25 =$

25) $4x - 5 + 15x + 3x^2 =$

26) $-7x^2 - 11x - 9x =$

27) $10x - 9x^2 - 3x^2 - 7 =$

28) $13 + 3x^2 - 9x^2 - 21x =$

29) $22x + 10x^2 - 15x + 17 =$

30) $4x^2 + 25x + 21x^2 =$

31) $29 - 12x^2 - 23x - 4x^2 =$

32) $22x - 19x - 9x^2 + 30 =$

PSSA Subject Test Mathematics Grade 7

The Distributive Property

✎ Use the distributive property to simply each expression.

1) $4(1 + 2x) =$

2) $2(4 + 7x) =$

3) $3(4x − 4) =$

4) $(2x − 5)(−6) =$

5) $(− 3)(x + 6) =$

6) $(4 + 3x)2 =$

7) $(−5)(8 − 3x) =$

8) $−(−5 − 7x) =$

9) $(−6x + 3)(−3) =$

10) $(−4)(x − 7) =$

11) $−(5 − 3x) =$

12) $3(9 + 4x) =$

13) $6(4 + 3x) =$

14) $(−5x + 3)2 =$

15) $(5 − 8x)(−3) =$

16) $(−12)(3x + 3) =$

17) $(5 − 3x)6 =$

18) $4(2 + 6x) =$

19) $8(7x − 3) =$

20) $(−2x + 3)4 =$

21) $(7 − 5x)(−9) =$

22) $(−10)(x − 8) =$

23) $(11 − 4x)3 =$

24) $(−6)(10x − 4) =$

25) $(3 − 9x)(−7) =$

26) $(−9)(x + 9) =$

27) $(−3 + 5x)(−7) =$

28) $(−5)(8 − 10x) =$

29) $12(4x − 8) =$

30) $(−10x + 13)(−3) =$

31) $(−8)(3x − 2) + 4(x + 5) =$

32) $(−8)(x + 4) − (6 + 5x) =$

WWW.MathNotion.Com

PSSA Subject Test Mathematics Grade 7

Evaluating One Variable Expressions

✏️ **Evaluate each expression using the value given.**

1) $8 - x$, $x = 5$

2) $x - 9$, $x = 5$

3) $5x + 4$, $x = 3$

4) $x - 13$, $x = -4$

5) $12 - x$, $x = 4$

6) $x + 2$, $x = 6$

7) $4x + 8$, $x = 3$

8) $x + (-7)$, $x = -8$

9) $4x + 5$, $x = 2$

10) $3x + 9$, $x = -2$

11) $15 + 3x - 7$, $x = 2$

12) $17 - 3x$, $x = 3$

13) $8x - 9$, $x = 4$

14) $5x + 4$, $x = -3$

15) $10x + 5$, $x = 3$

16) $14 - 4x$, $x = -6$

17) $3(5x + 3)$, $x = 9$

18) $4(-3x - 6)$, $x = 3$

19) $7x - 2x + 12$, $x = 4$

20) $(5x + 6) \div 2$, $x = 8$

21) $(x + 18) \div 10$, $x = 12$

22) $5x - 12 + 3x$, $x = -3$

23) $(6 - 4x)(-3)$, $x = -4$

24) $9x^2 + 3x - 6$, $x = 2$

25) $x^2 - 10x$, $x = -5$

26) $3x(7 - 2x)$, $x = 2$

27) $12x + 6 - 2x^2$, $x = -4$

28) $(-3)(4x - 8 + 3x)$, $x = 3$

29) $(-6) + \frac{x}{4} + 3x$, $x = 16$

30) $(-6) + \frac{x}{5}$, $x = 35$

31) $\left(-\frac{45}{x}\right) - 7 + 2x$, $x = 9$

32) $\left(-\frac{21}{x}\right) - 12 + 4x$, $x = 7$

WWW.MathNotion.Com

PSSA Subject Test Mathematics Grade 7

Evaluating Two Variables Expressions

✏️ **Evaluate each expression using the values given.**

1) $2x - 4y$,

   $x = 4, y = 1$

2) $3x + 5y$,

   $x = -2, y = 2$

3) $-7a + 4b$,

   $a = 2, b = 4$

4) $3x + 5 - y$,

   $x = 5, y = 6$

5) $3z + 12 - 2k$,

   $z = 5, k = 6$

6) $6(-x - 3y)$,

   $x = 5, y = -2$

7) $5a + 3b$,

   $a = 3, b = 4$

8) $7x \div 3y$,

   $x = 3, y = 7$

9) $2x + 15 + 5y$,

   $x = -3, y = 1$

10) $5a - (18 - b)$,

    $a = 2, b = 8$

11) $2z + 20 + 5k$,

    $z = -6, k = 5$

12) $xy + 10 + 4x$,

    $x = 3, y = 5$

13) $2x + 4y - 8 + 5$,

    $x = 5, y = 2$

14) $\left(-\frac{24}{x}\right) + 3 + 2y$,

    $x = 4, y = 6$

15) $(-3)(-3a - 3b)$,

    $a = 4, b = 5$

16) $12 + 4x - 7 - y$,

    $x = 3, y = 5$

17) $11x + 5 - 8y + 6$,

    $x = 5, y = 2$

18) $10 + 2(-4x - 5y)$,

    $x = 5, y = 4$

19) $5x + 13 + 6y$,

    $x = 5, y = 6$

20) $10a - (7a + 3b) - 11$,

    $a = 3, b = 8$

WWW.MathNotion.Com

PSSA Subject Test Mathematics Grade 7

Combining like Terms

✏️ Simplify each expression.

1) $11x + 3x + 6 =$

2) $8(2x - 6) =$

3) $18x - 7x + 11 =$

4) $(-4)(6x - 7) =$

5) $22x - 10x - 5 =$

6) $32x - 13 + 8x =$

7) $15 - (8x - 11) =$

8) $-24x + 17 - 11x =$

9) $12x - 8 - 6x + 9 =$

10) $21x + 5 - 36 + 12x =$

11) $28x + 3x - 11 =$

12) $(-3x + 4)5 =$

13) $2 + 4x + 9x - 8 =$

14) $6(2x - 5x) - 4 =$

15) $4(5x + 11) + 3x =$

16) $x - 14 - 11x =$

17) $5(10 + 9x) - 8x =$

18) $42x + 17 - 23x =$

19) $(-7x) + 19 + 20x =$

20) $(-7x) - 33 + 29x =$

21) $4(5x + 3) - 19x =$

22) $5(6 - 2x) - 15x =$

23) $-24x + (11 - 18x) =$

24) $(-9) - (6)(7x + 3) =$

25) $(-1)(8x - 10) - 21x =$

26) $-36x + 14 + 27x - 5x =$

27) $3(-13x + 6) - 17x =$

28) $-5x - 42 + 32x =$

29) $37x - 19x + 15 - 9x =$

30) $3(5x + 7x) - 31 =$

31) $14 - 6x - 15 - 9x =$

32) $-2(-5x - 7x) + 27x =$

WWW.MathNotion.Com

PSSA Subject Test Mathematics Grade 7

Answers of Worksheets

Translate Phrases into an Algebraic Statement

1) $9x$
2) $y - 11$
3) $\frac{19}{x}$
4) $38 - y$
5) $y + 40$
6) 6^2
7) x^5
8) $6 + x$
9) $57 - y$
10) $\frac{9}{x}$
11) $\frac{x^2}{25}$
12) $x - 6 = 19$
13) $10a - b^2$
14) $41 - ab$

Simplifying Variable Expressions

1) $3x + 15$
2) $-28x + 20$
3) $5x + 5$
4) $-8x^2 - 4$
5) $13x^2 + 10$
6) $18x^2 + 7x$
7) $-9x^2 + 4x$
8) $4x^2 - 10x$
9) $-22x + 21$
10) $68x - 12$
11) $-18x - 71$
12) $-11x^2 + 5x$
13) $-6x + 12$
14) $x - 4$
15) $22x - 5$
16) $14x - 22$
17) $27x - 16$
18) $25x + 28$
19) $-23x - 20$
20) $-3x^2 - 14x$
21) $-20x^2 + 27x$
22) $-20x^2 - 15x$
23) $4x^2 + 25x - 19$
24) $6x^2 - 66x + 25$
25) $3x^2 + 19x - 5$
26) $-7x^2 - 20x$
27) $-12x^2 + 10x - 7$
28) $-6x^2 - 21x + 13$
29) $10x^2 + 7x + 17$
30) $25x^2 + 25x$
31) $-16x^2 - 23x + 29$
32) $-9x^2 + 3x + 30$

The Distributive Property

1) $8x + 4$
2) $14x + 8$
3) $12x - 12$
4) $-12x + 30$
5) $-3x - 18$
6) $6x + 8$
7) $15x - 40$
8) $7x + 5$
9) $18x - 9$
10) $-4x + 28$
11) $3x - 5$
12) $12x + 27$
13) $18x + 24$
14) $-10x + 6$
15) $24x - 15$
16) $-36x - 36$
17) $-18x + 30$
18) $24x + 8$
19) $56x - 24$
20) $-8x + 12$
21) $45x - 63$
22) $-10x + 80$
23) $-12x + 33$
24) $-60x + 24$
25) $63x - 21$
26) $-9x - 81$
27) $-35x + 21$
28) $50x - 40$
29) $48x - 96$
30) $30x - 39$
31) $-20x + 36$
32) $-13x - 38$

Evaluating One Variables

1) 3
2) −4
3) 19
4) −17
5) 8
6) 8
7) 20
8) −15
9) 13
10) 3
11) 14
12) 8

WWW.MathNotion.Com

PSSA Subject Test Mathematics Grade 7

13) 23
14) −11
15) 35
16) 38
17) 144
18) −60
19) 32
20) 23
21) 3
22) −36
23) −66
24) 36
25) 75
26) 18
27) −74
28) −39
29) 46
30) 1
31) 6
32) 13

Evaluating Two Variables

1) 4
2) 4
3) 2
4) 14
5) 15
6) 6
7) 27
8) 1
9) 14
10) 0
11) 33
12) 37
13) 15
14) 9
15) 81
16) 12
17) 50
18) −70
19) 74
20) −26

Combining like Terms

1) $14x + 6$
2) $16x − 48$
3) $11x + 11$
4) $−24x + 28$
5) $12x − 5$
6) $40x − 13$
7) $−8x + 26$
8) $−35x + 17$
9) $6x + 1$
10) $33x − 31$
11) $31x − 11$
12) $−15x + 20$
13) $13x − 6$
14) $−18x − 4$
15) $23x + 44$
16) $−10x − 14$
17) $37x + 50$
18) $19x + 17$
19) $13x + 19$
20) $22x − 33$
21) $x + 12$
22) $−25x + 30$
23) $−42x + 11$
24) $−42x − 27$
25) $−29x + 10$
26) $−14x + 14$
27) $−56x + 18$
28) $27x − 42$
29) $9x + 15$
30) $36x − 31$
31) $−15x − 1$
32) $51x$

PSSA Subject Test Mathematics Grade 7

Chapter 7 :
Equations and Inequalities

Topics that you will practice in this chapter:

- ✓ One-Step Equations
- ✓ Multi-Step Equations
- ✓ Graphing Single-Variable Inequalities
- ✓ One-Step Inequalities
- ✓ Multi-Step Inequalities
- ✓ Systems of Equations
- ✓ Systems of Equations Word Problems

"Life is a math equation. In order to gain the most, you have to know how to convert negatives into positives." – Anonymous

PSSA Subject Test Mathematics Grade 7

One–Step Equations

✏️ **Find the answer for each equation.**

1) $3x = 90$, $x =$ _____

2) $5x = 35$, $x =$ _____

3) $6x = 24$, $x =$ _____

4) $24x = 144$, $x =$ _____

5) $x + 15 = 20$, $x =$ _____

6) $x - 7 = 4$, $x =$ _____

7) $x - 9 = 2$, $x =$ _____

8) $x + 15 = 23$, $x =$ _____

9) $x - 4 = 13$, $x =$ _____

10) $12 = 16 + x$, $x =$ _____

11) $x - 10 = 2$, $x =$ _____

12) $5 - x = -11$, $x =$ _____

13) $28 = -6 + x$, $x =$ _____

14) $x - 20 = -35$, $x =$ _____

15) $x + 14 = -4$, $x =$ _____

16) $14 = 28 - x$, $x =$ _____

17) $7 + x = -7$, $x =$ _____

18) $x - 16 = 4$, $x =$ _____

19) $30 = x - 15$, $x =$ _____

20) $x - 5 = -18$, $x =$ _____

21) $x - 10 = 24$, $x =$ _____

22) $x - 20 = -25$, $x =$ _____

23) $x - 17 = 30$, $x =$ _____

24) $-70 = x - 28$, $x =$ _____

25) $x - 9 = 13$, $x =$ _____

26) $36 = 4x$, $x =$ _____

27) $x - 35 = 25$, $x =$ _____

28) $x - 25 = 10$, $x =$ _____

29) $70 - x = 16$, $x =$ _____

30) $x - 10 = 14$, $x =$ _____

31) $17 - x = -13$, $x =$ _____

32) $x - 9 = -30$, $x =$ _____

WWW.MathNotion.Com

PSSA Subject Test Mathematics Grade 7

Multi–Step Equations

✏️ Find the answer for each equation.

1) $3x + 3 = 9$

2) $-x + 5 = 12$

3) $4x - 8 = 8$

4) $-(3 - x) = 5$

5) $4x - 8 = 16$

6) $12x - 15 = 9$

7) $2x - 18 = 2$

8) $4x + 8 = 16$

9) $24x + 27 = 75$

10) $-14(3 + x) = 14$

11) $-3(2 + x) = 6$

12) $12 = -(x - 7)$

13) $3(3 - x) = 30$

14) $-15 = -(3x + 6)$

15) $40(3 + x) = 40$

16) $5(x - 10) = 25$

17) $-18 = x + 8x$

18) $3x + 25 = -2x - 10$

19) $7(6 + 3x) = -63$

20) $18 - 3x = -4 - 5x$

21) $4 - 6x = 36 + 2x$

22) $15 + 15x = -5 + 5x$

23) $42 = (-6x) - 7 + 7$

24) $21 = 3x - 21 + 4x$

25) $-18 = -6x - 9 + 3x$

26) $5x - 15 = -29 + 6x$

27) $7x - 18 = 4x + 3$

28) $-7 - 4x = 5(4 - x)$

29) $x - 5 = -5(-3 - x)$

30) $13x - 68 = 15x - 102$

31) $-5x - 3 = -3(9 + 3x)$

32) $-2x - 15 = 6x + 17$

WWW.MathNotion.Com

Graphing Single–Variable Inequalities

✎ **Draw a graph for each inequality.**

1) $x > -1$

2) $x \leq 2$

3) $x \geq 0$

4) $x < -3$

5) $x < \frac{1}{2}$

6) $x \leq -2$

7) $x \leq 3$

8) $x \geq -\frac{7}{2}$

One–Step Inequalities

✎ Find the answer for each inequality and graph it.

1) $x + 4 \geq 4$

2) $x - 5 \leq 2$

3) $5x > 35$

4) $9 + x \leq 11$

5) $x - 5 < -9$

6) $9x \geq 72$

7) $9x \leq 27$

8) $x + 19 > 16$

PSSA Subject Test Mathematics Grade 7

Multi-Step Inequalities

✎ **Calculate each inequality.**

1) $x - 3 \leq 7$

2) $8 - x \leq 8$

3) $3x - 9 \leq 9$

4) $4x - 4 \geq 8$

5) $x - 7 \geq 1$

6) $5x - 15 \leq 5$

7) $6x - 8 \leq 4$

8) $-11 + 6x \leq 12$

9) $4(x - 4) \leq 16$

10) $3x - 10 \leq 11$

11) $5x - 25 < 25$

12) $9x - 5 < 22$

13) $20 - 7x \geq -15$

14) $33 + 6x < 45$

15) $8 + 8x \geq 96$

16) $7 + 3x < 13$

17) $4x - 3 < 9$

18) $5(2 - 2x) \geq -30$

19) $-(7 + 6x) < 29$

20) $12 - 8x \geq -20$

21) $-4(x - 6) > 24$

22) $\dfrac{3x + 9}{6} \leq 10$

23) $\dfrac{4x - 10}{3} \leq 2$

24) $\dfrac{2x - 8}{3} > 2$

25) $8 + \dfrac{x}{6} < 9$

26) $\dfrac{9x}{7} - 4 < 5$

27) $\dfrac{15x + 45}{15} > 1$

28) $16 + \dfrac{x}{4} < 6$

WWW.MathNotion.Com

PSSA Subject Test Mathematics Grade 7

Systems of Equations

✏️ **Calculate each system of equations.**

1) $-x + y = 2$ $x = \underline{}$
    $-4x + 2y = 6$ $y = \underline{}$

2) $-15x + 3y = -9$ $x = \underline{}$
    $9x - 16y = 48$ $y = \underline{}$

3) $y = -7$ $x = \underline{}$
    $6x + 5y = 7$ $y = \underline{}$

4) $3y = -9x + 15$ $x = \underline{}$
    $5x - 4y = -3$ $y = \underline{}$

5) $10x - 9y = -13$ $x = \underline{}$
    $-5x + 3y = 11$ $y = \underline{}$

6) $-12x - 16y = 20$ $x = \underline{}$
    $6x - 12y = 30$ $y = \underline{}$

7) $5x - 14y = -23$ $x = \underline{}$
    $-18x + 21y = 24$ $y = \underline{}$

8) $15x - 21y = -6$ $x = \underline{}$
    $2x - 3y = -2$ $y = \underline{}$

9) $-x + 3y = 3$ $x = \underline{}$
    $-14x + 16y = -10$ $y = \underline{}$

10) $x + 5y = 50$ $x = \underline{}$
    $3x + 10y = 80$ $y = \underline{}$

11) $6x - 7y = -8$ $x = \underline{}$
    $-x - 4y = -9$ $y = \underline{}$

12) $2x + 4y = -10$ $x = \underline{}$
    $2x - 8y = 14$ $y = \underline{}$

13) $4x + 3y = 12$ $x = \underline{}$
    $5x - 3y = 15$ $y = \underline{}$

14) $3x - 2y = 3$ $x = \underline{}$
    $7x - 8y = 22$ $y = \underline{}$

15) $3x + 2y = 5$ $x = \underline{}$
    $-10x - 4y = -14$ $y = \underline{}$

16) $10x + 7y = 1$ $x = \underline{}$
    $-5x - 7y = 24$ $y = \underline{}$

WWW.MathNotion.Com

PSSA Subject Test Mathematics Grade 7

Systems of Equations Word Problems

✎ **Find the answer for each word problem.**

1) Tickets to a movie cost $4 for adults and $3 for students. A group of friends purchased 8 tickets for $31.00. How many adults ticket did they buy? _____

2) At a store, Eva bought two shirts and five hats for $77.00. Nicole bought three same shirts and four same hats for $84.00. What is the price of each shirt? _____

3) A farmhouse shelters 18 animals, some are pigs, and some are ducks. Altogether there are 66 legs. How many pigs are there? _____

4) A class of 214 students went on a field trip. They took 36 vehicles, some cars and some buses. If each car holds 5 students and each bus hold 22 students, how many buses did they take? _____

5) A theater is selling tickets for a performance. Mr. Smith purchased 5 senior tickets and 3 child tickets for $105 for his friends and family. Mr. Jackson purchased 3 senior tickets and 5 child tickets for $79. What is the price of a senior ticket? $_____

6) The difference of two numbers is 10. Their sum is 20. What is the bigger number? $_____

7) The sum of the digits of a certain two-digit number is 7. Reversing its digits increase the number by 9. What is the number? _____

8) The difference of two numbers is 11. Their sum is 25. What are the numbers? _____

9) The length of a rectangle is 5 meters greater than 2 times the width. The perimeter of rectangle is 28 meters. What is the length of the rectangle? _____

10) Jim has 25 nickels and dimes totaling $1.80. How many nickels does he have? _____

WWW.MathNotion.Com

PSSA Subject Test Mathematics Grade 7

Answers of Worksheets

One–Step Equations

1) 30	9) 17	17) −14	25) 22
2) 7	10) −4	18) 20	26) 9
3) 4	11) 12	19) 45	27) 60
4) 6	12) 16	20) −13	28) 35
5) 5	13) 34	21) 34	29) 54
6) 11	14) −15	22) −5	30) 24
7) 11	15) −18	23) 47	31) 30
8) 8	16) 14	24) −42	32) −21

Multi–Step Equations

1) 2	9) 2	17) −2	25) 3
2) −7	10) −4	18) −7	26) 14
3) 4	11) −4	19) −5	27) 7
4) 8	12) −5	20) −11	28) 27
5) 6	13) −7	21) −4	29) −5
6) 2	14) 3	22) −2	30) 17
7) 10	15) −2	23) −7	31) −6
8) 2	16) 15	24) 6	32) −4

Graphing Single–Variable Inequalities

1) $x > -1$ (open circle at −1, shaded right)

2) $x \leq 2$ (closed circle at 2, shaded left)

3) $x \geq 0$ (closed circle at 0, shaded right)

4) $x > -3$ (open circle at −3, shaded right)

PSSA Subject Test Mathematics Grade 7

5)

6)

7)

8)

One–Step Inequalities

1)

2)

3)

4)

5)

6)

7)

8)

Multi-Step Inequalities

1) $x \leq 10$
2) $x \geq 0$
3) $x \leq 6$
4) $x \geq 3$
5) $x \geq 8$
6) $x \leq 4$
7) $x \leq 2$
8) $x \leq \frac{23}{6}$
9) $x \leq 8$
10) $x \leq 7$
11) $x < 10$
12) $x < 3$
13) $x \leq 5$
14) $x < 2$
15) $x \geq 11$
16) $x < 2$
17) $x < 3$
18) $x \leq 4$
19) $x > -6$
20) $x \leq 4$
21) $x < 0$
22) $x \leq 17$
23) $x \leq 4$
24) $x > 7$

PSSA Subject Test Mathematics Grade 7

25) $x < 6$ 26) $x < 7$ 27) $x > -2$ 28) $x < -40$

Systems of Equations

1) $x = -1, y = 1$ 7) $x = 1, y = 2$ 13) $x = 3, y = 0$
2) $x = 0, y = -3$ 8) $x = 8, y = 6$ 14) $x = -2, y = -\frac{9}{2}$
3) $x = 7$ 9) $x = 3, y = 2$ 15) $x = 1, y = 1$
4) $x = 1, y = 2$ 10) $x = -20, y = 14$ 16) $x = 5, y = -7$
5) $x = -4, y = -3$ 11) $x = 1, y = 2$
6) $x = 1, y = -2$ 12) $x = -1, y = -2$

Systems of Equations Word Problems

1) 7 5) $18 9) 11 meters
2) $16 6) 15 10) 14
3) 15 7) 34
4) 2 8) 18, 7

Chapter 8:
Linear Functions

Topics that you will practice in this chapter:

- ✓ Finding Slope
- ✓ Graphing Lines Using Line Equation
- ✓ Writing Linear Equations
- ✓ Graphing Linear Inequalities
- ✓ Finding Midpoint
- ✓ Finding Distance of Two Points

"Nature is written in mathematical language." – Galileo Galilei

Finding Slope

👉 **Find the slope of each line.**

1) $y = x + 8$

2) $y = -3x + 5$

3) $y = 2x + 12$

4) $y = -4x + 19$

5) $y = 11 + 6x$

6) $y = 7 - 5x$

7) $y = 8x + 19$

8) $y = -9x + 20$

9) $y = -7x + 4$

10) $y = 3x - 8$

11) $y = \frac{1}{3}x + 8$

12) $y = -\frac{4}{5}x + 9$

13) $-3x + 6y = 30$

14) $4x + 4y = 16$

15) $3y - x = 10$

16) $8y - x = 5$

👉 **Find the slope of the line through each pair of points.**

17) $(2, 3), (7, 10)$

18) $(-3, 5), (2, 15)$

19) $(5, -3), (1, 9)$

20) $(-5, -5), (10, 25)$

21) $(22, 3), (7, 18)$

22) $(-16, 8), (-7, 26)$

23) $(25, 11), (29, 19)$

24) $(26, -19), (14, 17)$

25) $(22, -13), (20, -11)$

26) $(19, 7), (15, -3)$

27) $(5, 7), (11, 19)$

28) $(52, -62), (40, 70)$

PSSA Subject Test Mathematics Grade 7

Graphing Lines Using Line Equation

✏️ Sketch the graph of each line.

1) $y = x - 2$

2) $y = -3x + 2$

3) $x + y = 0$

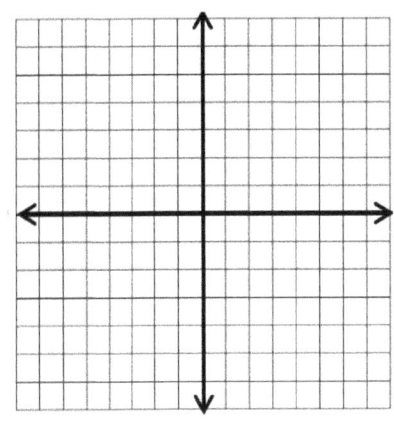

4) $x + y = -3$

5) $2x + 3y = -4$

6) $y - 3x + 6 = 0$

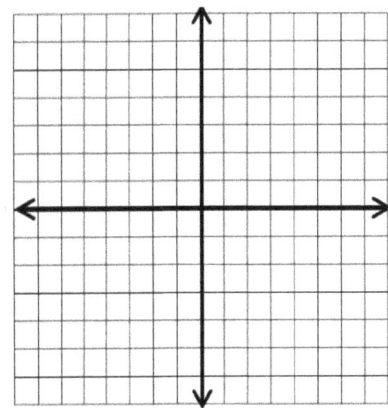

WWW.MathNotion.Com

PSSA Subject Test Mathematics Grade 7

Writing Linear Equations

✎ **Write the equation of the line through the given points.**

1) Through: $(2, -5), (3, 9)$

2) Through: $(-6, 3), (3, 12)$

3) Through: $(10, 7), (5, 27)$

4) Through: $(15, 11), (3, -1)$

5) Through: $(24, 17), (12, -7)$

6) Through: $(8, 29), (4, -7)$

7) Through: $(20, -16), (12, 0)$

8) Through: $(-3, 10), (2, -5)$

9) Through: $(-6, 17), (4, -3)$

10) Through: $(-8, 22), (5, -4)$

11) Through: $(9, 27), (3, -3)$

12) Through: $(11, 32), (9, 4)$

13) Through: $(-3, 13), (-4, 0)$

14) Through: $(-5, 5), (5, 15)$

15) Through: $(18, -32), (11, 3)$

16) Through: $(-4, 25), (4, -15)$

✎ **Find the answer for each problem.**

17) What is the equation of a line with slope 6 and intercept 12? _____

18) What is the equation of a line with slope −11 and intercept −4? _____

19) What is the equation of a line with slope −3 and passes through point $(5, 2)$? _____

20) What is the equation of a line with slope −5 and passes through point $(-2, -1)$? _____

21) The slope of a line is −10 and it passes through point $(-3, 0)$. What is the equation of the line? _____

22) The slope of a line is 8 and it passes through point $(0, 7)$. What is the equation of the line? _____

Graphing Linear Inequalities

✎ **Sketch the graph of each linear inequality.**

1) $y > 4x - 5$

2) $y < 2x + 4$

3) $y \leq -5x - 2$

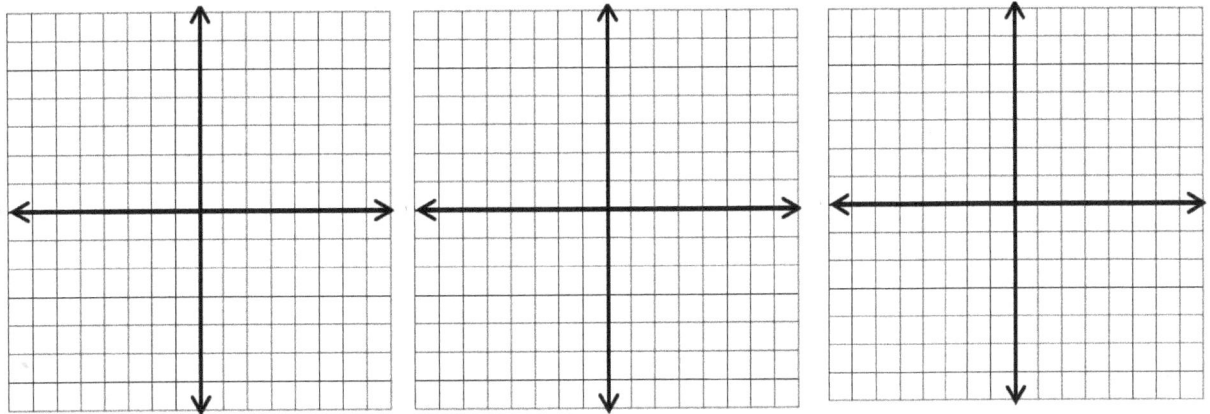

4) $4y \geq 12 + 4x$

5) $-12y < 3x - 24$

6) $5y \geq -15x + 10$

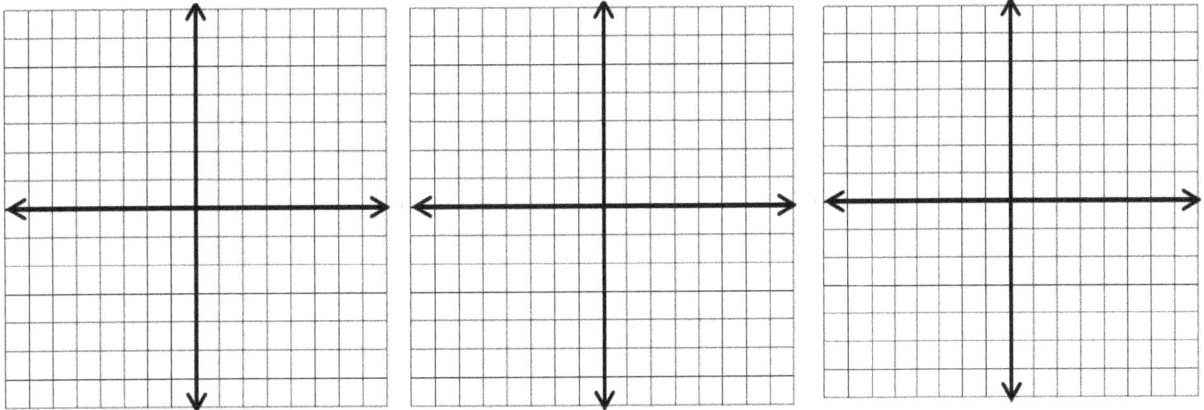

PSSA Subject Test Mathematics Grade 7

Finding Midpoint

✎ **Find the midpoint of the line segment with the given endpoints.**

1) $(-4, -3), (2, 3)$

2) $(9, 0), (-1, 8)$

3) $(9, -6), (3, 14)$

4) $(-10, -6), (0, 8)$

5) $(2, -5), (14, -15)$

6) $(-10, -3), (4, -13)$

7) $(8, 7), (-8, 13)$

8) $(-3, 6), (-9, 2)$

9) $(-4, 5), (16, -9)$

10) $(7, 14), (9, -2)$

11) $(-8, 6), (6, 6)$

12) $(10, 5), (-2, -3)$

13) $(-5, 12), (-3, 3)$

14) $(12, 7), (8, -2)$

15) $(10, 2), (-6, 14)$

16) $(-1, -2), (-7, 10)$

17) $(7, -7), (13, -13)$

18) $(-3, -8), (11, -4)$

19) $(5, -11), (-8, 9)$

20) $(14, -4), (16, 14)$

21) $(0, -5), (8, -1)$

22) $(3, 0), (-21, 18)$

23) $(17, -3), (-7, -5)$

24) $(26, -12), (6, 24)$

✎ **Find the answer for each problem.**

25) One endpoint of a line segment is $(-3, 7)$ and the midpoint of the line segment is $(-6, 9)$. What is the other endpoint? _____

26) One endpoint of a line segment is $(-3, 7)$ and the midpoint of the line segment is $(1, 5)$. What is the other endpoint? _____

27) One endpoint of a line segment is $(-10, -16)$ and the midpoint of the line segment is $(2, 9)$. What is the other endpoint? _____

WWW.MathNotion.Com

PSSA Subject Test Mathematics Grade 7

Finding Distance of Two Points

✏️ **Find the distance between each pair of points.**

1) $(6, 3), (-3, -9)$

2) $(5, 2), (-10, -6)$

3) $(8, 5), (8, 3)$

4) $(-8, -2), (2, 22)$

5) $(6, -7), (-3, -7)$

6) $(12, 0), (-9, -20)$

7) $(3, 20), (3, -5)$

8) $(10, 17), (5, 5)$

9) $(7, -2), (-4, -2)$

10) $(13, 4), (5, -2)$

11) $(11, 13), (5, 5)$

12) $(1, 4), (-23, -3)$

13) $(9, 8), (5, -4)$

14) $(-11, -4), (5, 8)$

15) $(-2, -6), (-2, -12)$

16) $(-1, -4), (23, 3)$

17) $(19, 3), (7, -6)$

18) $(-5, -2), (3, 4)$

19) $(2, 6), (2, -12)$

20) $(-4, -2), (8, -2)$

✏️ **Find the answer for each problem.**

21) Triangle ABC is a right triangle on the coordinate system and its vertices are $(-2, 5)$, $(-2, 1)$, and $(1, 1)$. What is the area of triangle ABC? _____

22) Three vertices of a triangle on a coordinate system are $(3, -6)$, $(-5, -12)$, and $(3, -18)$. What is the perimeter of the triangle? _____

23) Four vertices of a rectangle on a coordinate system are $(-2, 2)$, $(-2, 6)$, $(4, 2)$, and $(4, 6)$. What is its perimeter? _____

WWW.MathNotion.Com

PSSA Subject Test Mathematics Grade 7

Answers of Worksheets

Finding Slope

1) 1
2) −3
3) 2
4) −4
5) 6
6) −5
7) 8
8) −9
9) −7
10) 3
11) $\frac{1}{3}$
12) $-\frac{4}{5}$
13) $\frac{1}{2}$
14) −1
15) $\frac{1}{3}$
16) $\frac{1}{8}$
17) $\frac{7}{5}$
18) 2
19) −3
20) 2
21) −1
22) 2
23) 2
24) −3
25) −1
26) $\frac{5}{2}$
27) 2
28) −11

Graphing Lines Using Line Equation

1) $y = x - 2$

2) $y = -3x + 2$

3) $x + y = 0$

4) $x + y = -3$

5) $2x + 3y = -4$

6) $y - 3x + 6 = 0$

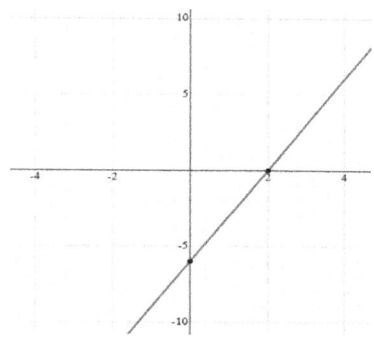

PSSA Subject Test Mathematics Grade 7

Writing Linear Equations

1) $y = 14x - 33$
2) $y = x + 9$
3) $y = -4x + 47$
4) $y = x - 4$
5) $y = 2x - 31$
6) $y = 9x - 43$
7) $y = -2x + 24$
8) $y = -3x + 1$
9) $y = -2x + 5$
10) $y = -2x + 6$
11) $y = 5x - 18$
12) $y = 14x - 122$
13) $y = 13x + 52$
14) $y = x + 10$
15) $y = -5x + 58$
16) $y = -5x + 5$
17) $y = 6x + 12$
18) $y = -11x - 4$
19) $y = -3x + 17$
20) $y = -5x - 11$
21) $y = -10x - 30$
22) $y = 8x + 7$

Graphing Linear Inequalities

1) $y > 4x - 5$

2) $y < 2x + 4$

3) $y \leq -5x - 2$

4) $4y \geq 12 + 4x$

5) $-12y < 3x - 24$

6) $5y \geq -15x + 10$

Finding Midpoint

1) $(-1, 0)$
2) $(4, 4)$
3) $(6, 4)$
4) $(-5, 1)$
5) $(8, -10)$
6) $(-3, -8)$
7) $(0, 10)$
8) $(-6, 4)$
9) $(6, -2)$
10) $(8, 6)$
11) $(-1, 6)$
12) $(4, 1)$
13) $(-4, 7.5)$
14) $(10, 2.5)$
15) $(2, 8)$
16) $(-4, 4)$
17) $(10, -10)$
18) $(4, -6)$

PSSA Subject Test Mathematics Grade 7

19) $(-1.5, -1)$
20) $(15, 5)$
21) $(4, -3)$

22) $(-9, 9)$
23) $(5, -4)$
24) $(16, 6)$

25) $(-9, 11)$
26) $(5, 3)$
27) $(14, 34)$

Finding Distance of Two Points

1) 15
2) 17
3) 2
4) 26
5) 9
6) 29
7) 25
8) 13

9) 11
10) 10
11) 10
12) 25
13) $4\sqrt{10}$
14) 20
15) 6
16) 25

17) 15
18) 10
19) 18
20) 12
21) 6 square units
22) 32 units
23) 20 units

Chapter 9:
Transformations

Topics that you will learn in this chapter:

- ✓ Translations
- ✓ Reflections
- ✓ Rotations
- ✓ Dilations
- ✓ Coordinate of Vertices.

"Nothing gets transformed in your life until your mind is transformed."

Ifeanyi Enoch Onuoha

PSSA Subject Test Mathematics Grade 7

Translations

✎ **Graph the image of the figure using the transformation given.**

1) Translation: 5 units right and 2 units down

2) Translation: 1 units left and 3 units down

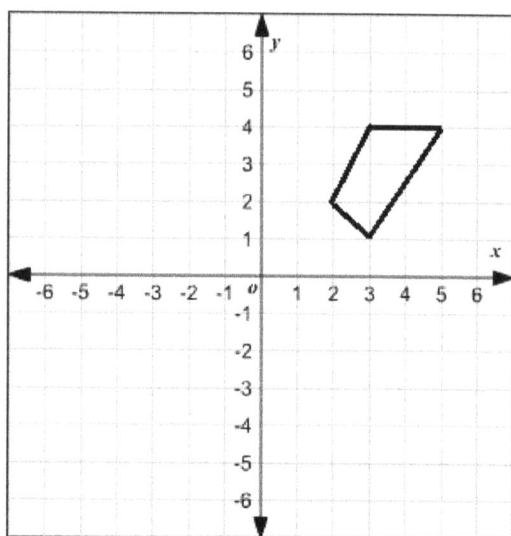

✎ **Write a rule to describe each transformation.**

3)

4)

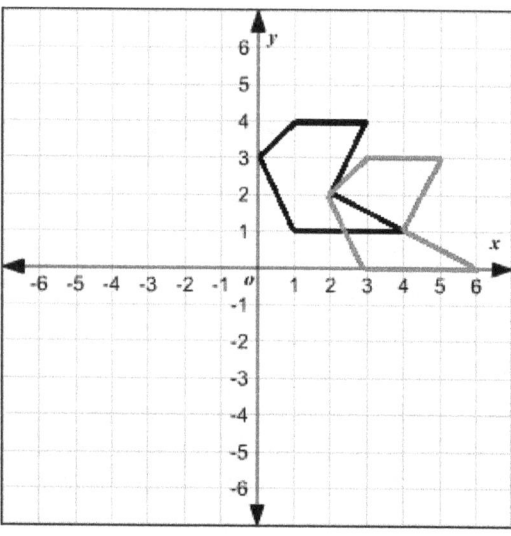

Reflections

✎ **Graph the image of the figure using the transformation given.**

1) Reflection across $x = 3$

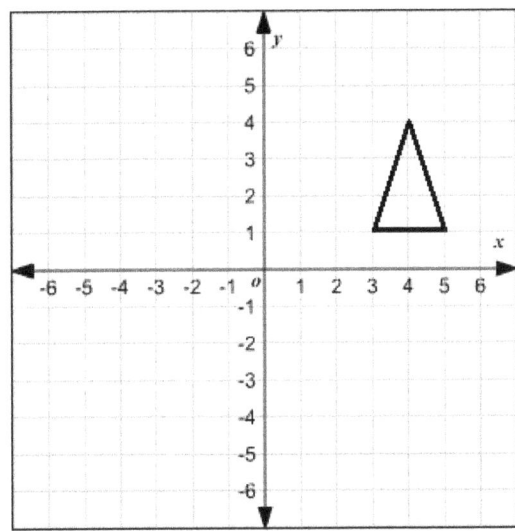

2) Reflection across $y = x$

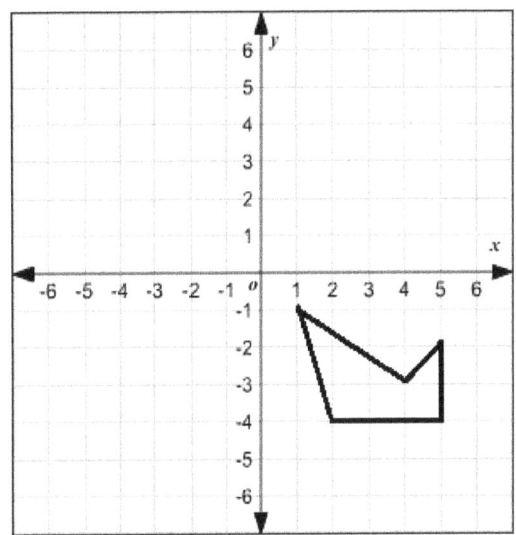

3) Reflection across $y = -2$

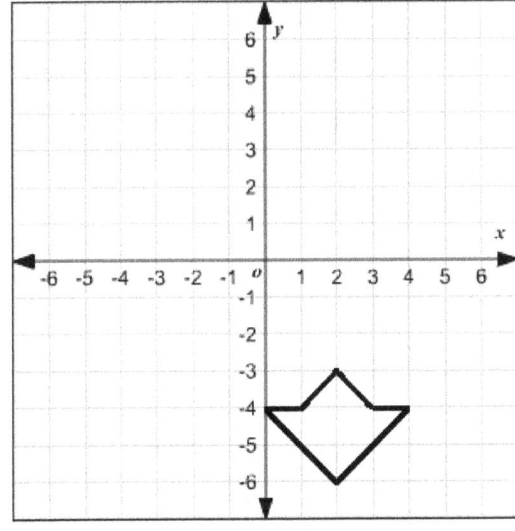

4) Reflection across y axis

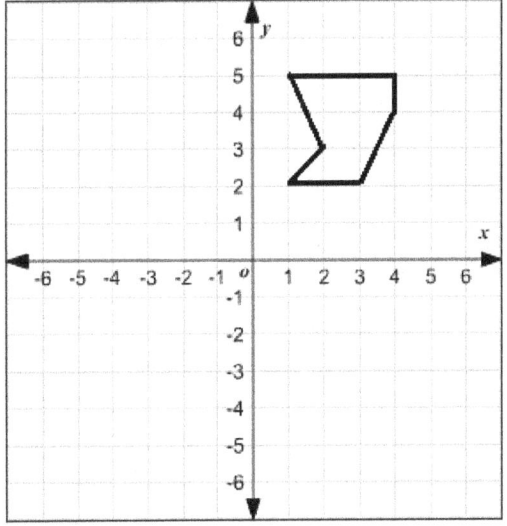

PSSA Subject Test Mathematics Grade 7

✎ **Write a rule to describe each transformation.**

5)

6)

7)

8)

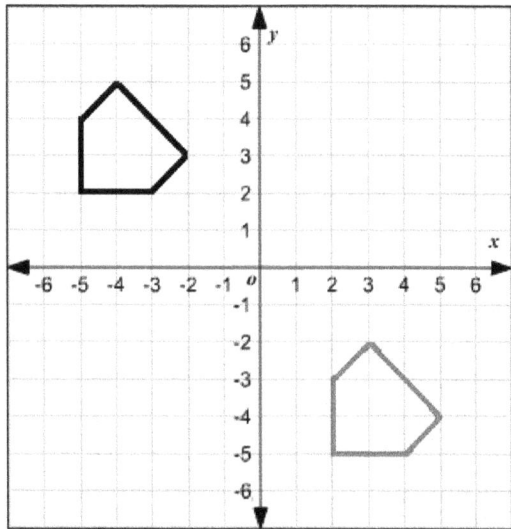

PSSA Subject Test Mathematics Grade 7

Rotations

✎ **Graph the image of the figure using the transformation given.**

1) Rotation 90° counterclockwise about the origin

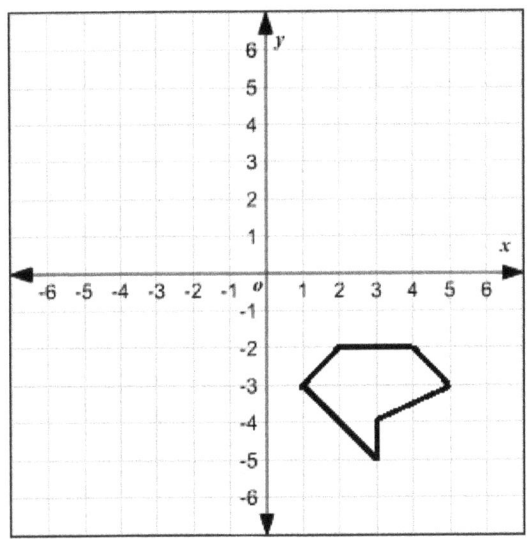

2) Rotation 270° counterclockwise about the origin

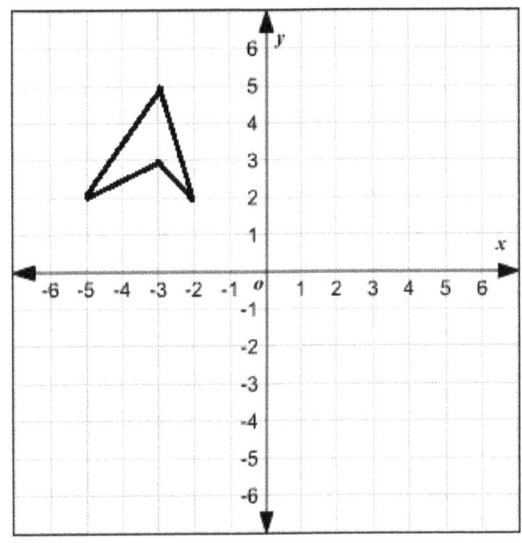

3) Rotation 270° clockwise about the origin

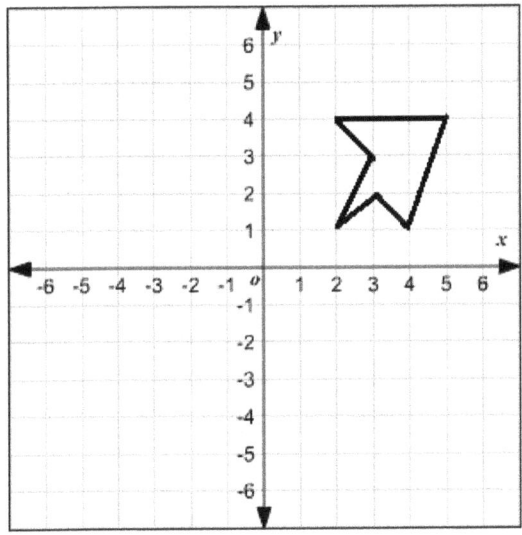

4) Rotation 180° counterclockwise about the origin

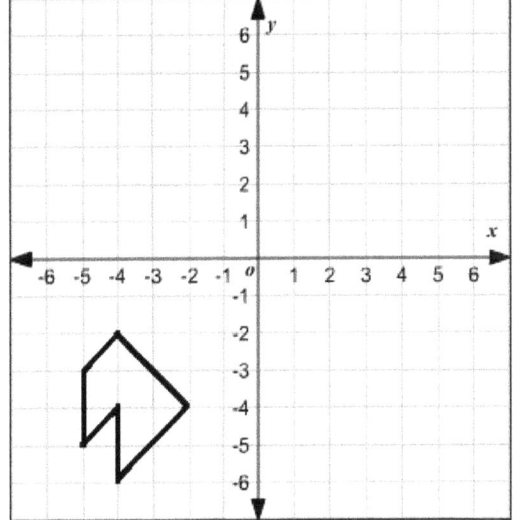

PSSA Subject Test Mathematics Grade 7

🖎 **Write a rule to describe each transformation.**

5)

6)

7)

8)

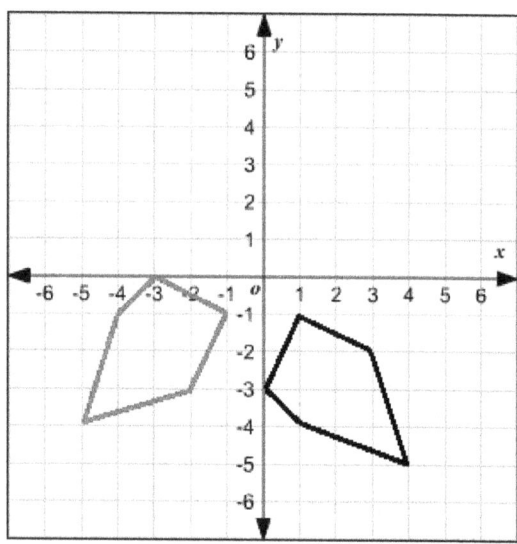

WWW.MathNotion.Com

PSSA Subject Test Mathematics Grade 7

Dilations

✎ Determine whether the dilation from figure M to figure N is a reduction or an enlargement. Then find the scale factor and the missing length.

1)

2)

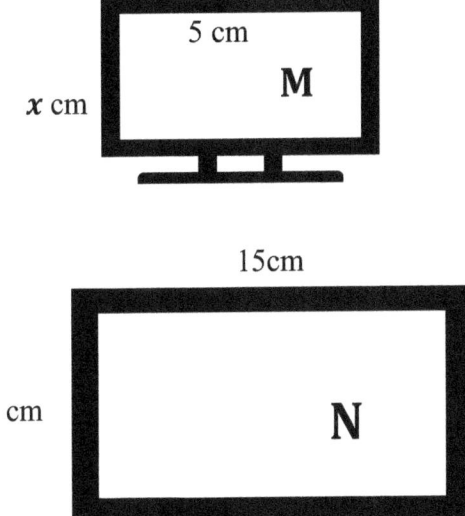

✎ Draw a dilation of the figure using the given scale factor.

3) $k = \frac{1}{2}$

4) $k = 3$

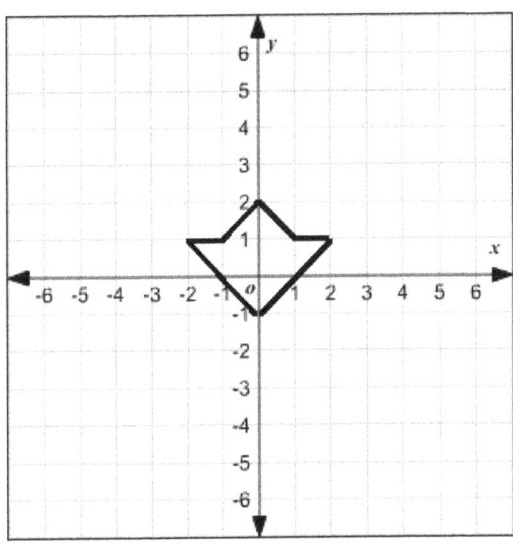

WWW.MathNotion.Com

PSSA Subject Test Mathematics Grade 7

Coordinates of Vertices

✏ **Calculate the new coordinates after the given transformations.**

1) Translate: 1-unit right and 2 units down.

   A(−1, 0), B(1, −3), C(2, 1)

2) Rotation: 270° clockwise about the origin.

   D(−1, 1), E(−5, 6), F(−7, 3), G(−3, 10)

3) Rotation: 180° counterclockwise about the origin.

   P(3, 0), Q(5, 2), R(−1, 4), S(2, −7)

4) Rotation: 90° clockwise about the origin.

   J(−2, 5), K(−5, 0), L(3, −6)

5) Reflection: over the y axis.

   C(−2, −6), D(4, −3), W(1, −7), Y(5, 1)

6) Reflection: across the line $y = -x$.

   A(3, −2), B(8, −4), C(6, −6), D(1, −5)

7) Reflection: across the line $y = -2$.

   K(−1, 1), L(−4, 2), M(4, −1), N(2, 3)

8) Dilate: Reduction by scale factor $\frac{1}{3}$.

   A(6, 3), B(−12, 0), C(−9, 6)

9) Dilate: Enlargement by scale factor 2.

   F(−1, 4), G(−3, 0), H(3, 2)

PSSA Subject Test Mathematics Grade 7

Answers of Worksheets

Translations

1)

2)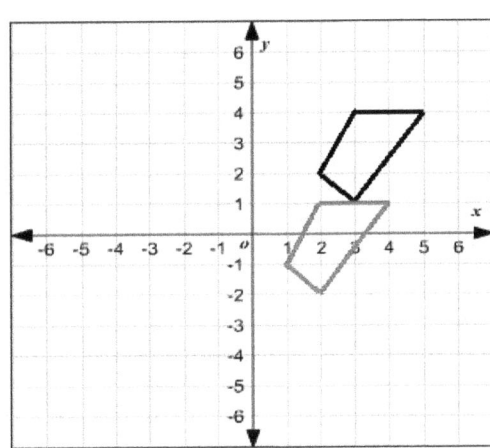

3) Translation: : 1 units left and 4 units down

4) Translation: 2 units right and 1 unit down

Reflections

1)

2)

3)

4)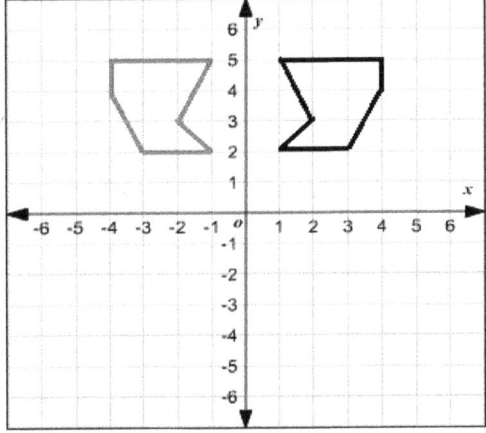

WWW.MathNotion.Com

5) Reflection across the x = 0 (y axis)

6) Reflection across the y = 1

7) Reflection against the origin

8) Reflection across the y = x

Rotations

1)

2)

3)

4)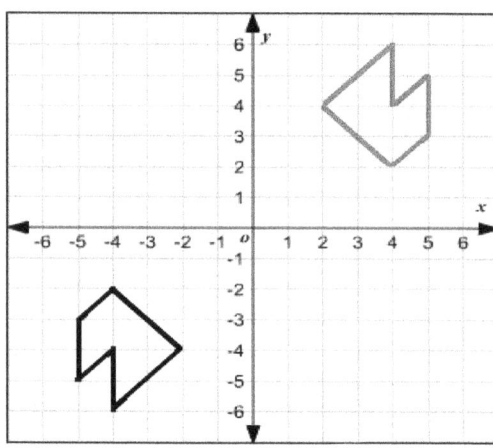

5) Rotation 90° clockwise about the origin

6) Rotation 180° about the origin

7) Rotation 270° counter clockwise about the origin

8) Rotation 90° clockwise about the origin

Dilations

1) Reduction, $k = \frac{3}{2}$, $x = 4\ cm$

2) Enlargement, $k = \frac{1}{3}$, $x = 3\ cm$

PSSA Subject Test Mathematics Grade 7

3)

4)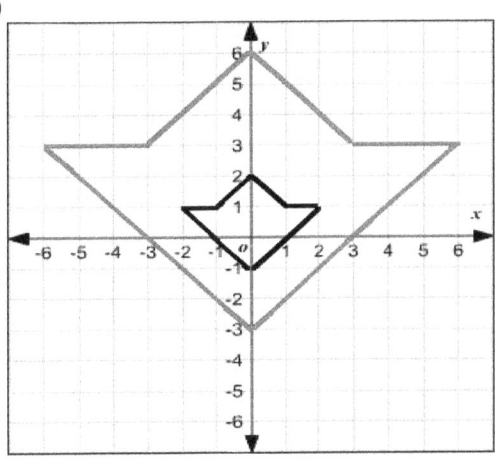

Coordinate of Vertices

1) $A'(0, -2), B'(2, -5), C'(3, -1)$
2) $D'(-1, -1), E'(-6, -5), F'(-3, -7), G'(-10, -3)$
3) $P'(-3, 0), Q'(-5, -2), R'(1, -4), S'(-2, 7)$
4) $J'(5, 2), K'(0, 5), L'(-6, -3)$
5) $C'(2, -6), D'(-4, -3), W'(-1, -7), Y'(-5, 1)$
6) $A'(2, -3), B'(4, -8), C'(6, -6), D'(5, -1)$
7) $K'(-1, -5), L'(-4, -6), M'(4, -3), N'(2, -7)$
8) $A'(2, 1), B'(-4, 0), C'(-3, 2)$
9) $F'(-2, 8), G'(-6, 0), H'(6, 4)$

PSSA Subject Test Mathematics Grade 7

Chapter 10:
Geometry and Solid Figures

Topics that you will practice in this chapter:

- ✓ Angles
- ✓ Pythagorean Relationship
- ✓ Triangles
- ✓ Polygons
- ✓ Trapezoids
- ✓ Circles
- ✓ Cubes
- ✓ Rectangular Prism
- ✓ Cylinder
- ✓ Pyramids and Cone

Mathematics is, as it were, a sensuous logic, and relates to philosophy as do the arts, music, and plastic art to poetry. — K. Shegel

PSSA Subject Test Mathematics Grade 7

Angles

✏️ **What is the value of *x* in the following figures?**

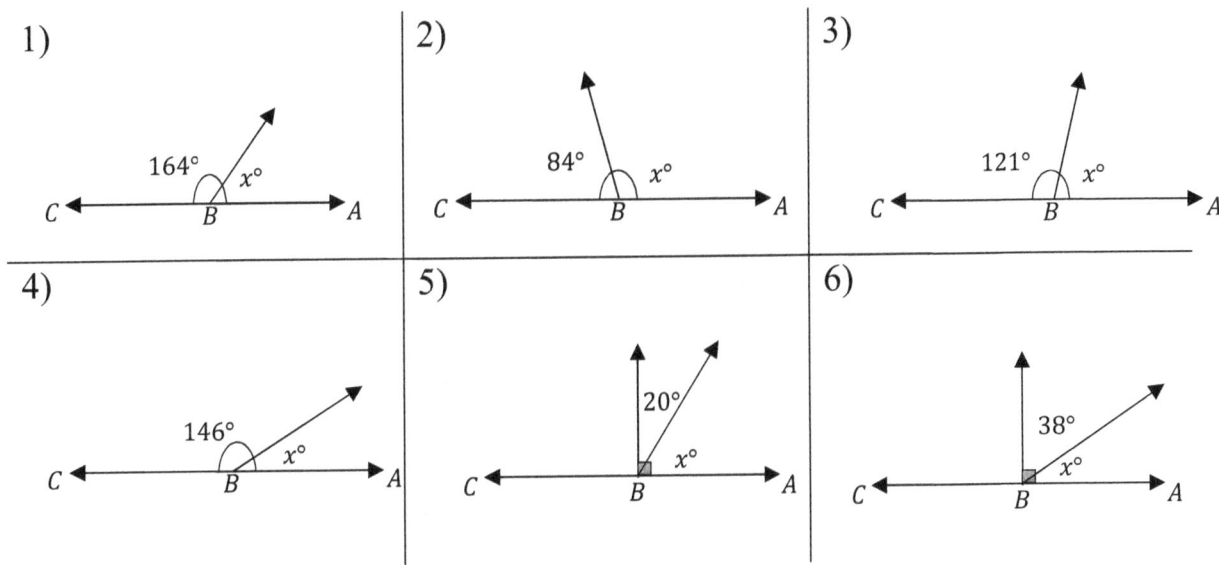

✏️ **Calculate.**

7) Two supplement angles have equal measures. What is the measure of each angle? _____

8) The measure of an angle is seven fifth the measure of its supplement. What is the measure of the angle? _____

9) Two angles are complementary and the measure of one angle is 24 less than the other. What is the measure of the smaller angle? _____

10) Two angles are complementary. The measure of one angle is one fifth the measure of the other. What is the measure of the bigger angle? _____

11) Two supplementary angles are given. The measure of one angle is 40° less than the measure of the other. What does the smaller angle measure? _____

WWW.MathNotion.Com

PSSA Subject Test Mathematics Grade 7

Pythagorean Relationship

✎ Do the following lengths form a right triangle?

1)

2)

3)

4)

5)

6)

7)

8)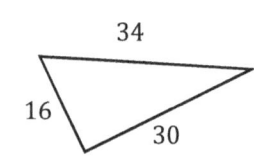

✎ Find the missing side?

9)

10)

11)

12)

13)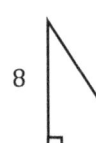

14)

15)

16)

WWW.MathNotion.Com

Triangles

✏ **Find the measure of the unknown angle in each triangle.**

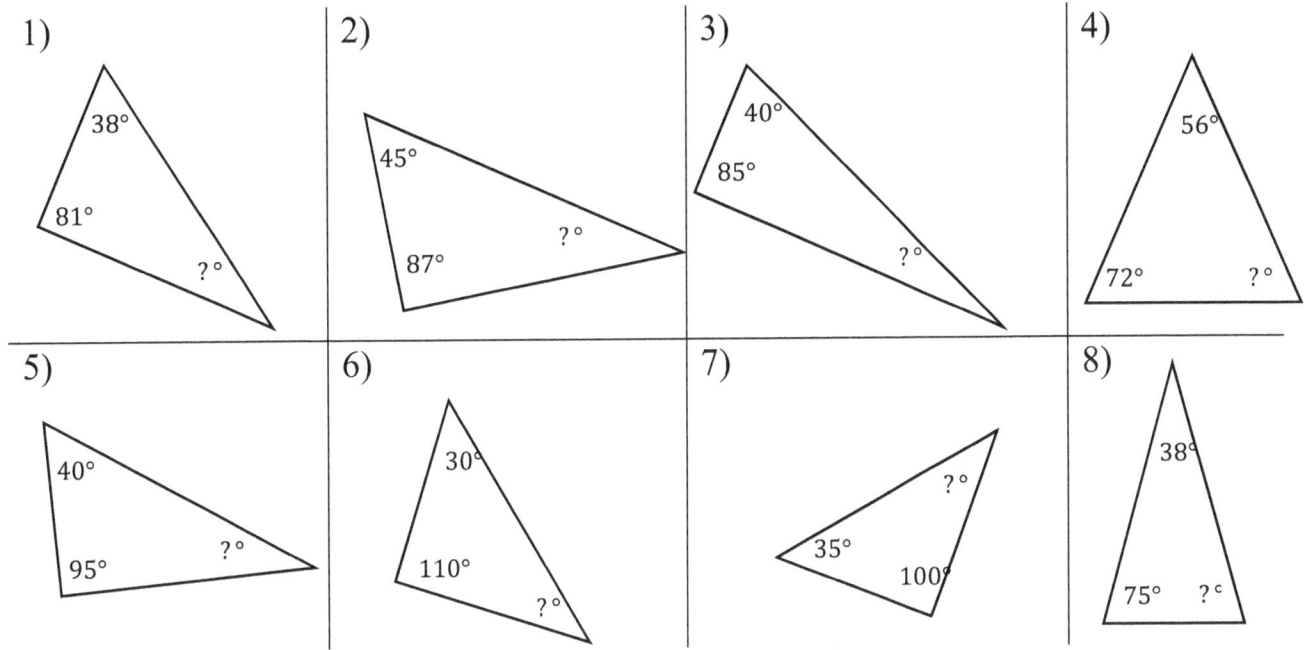

✏ **Find area of each triangle.**

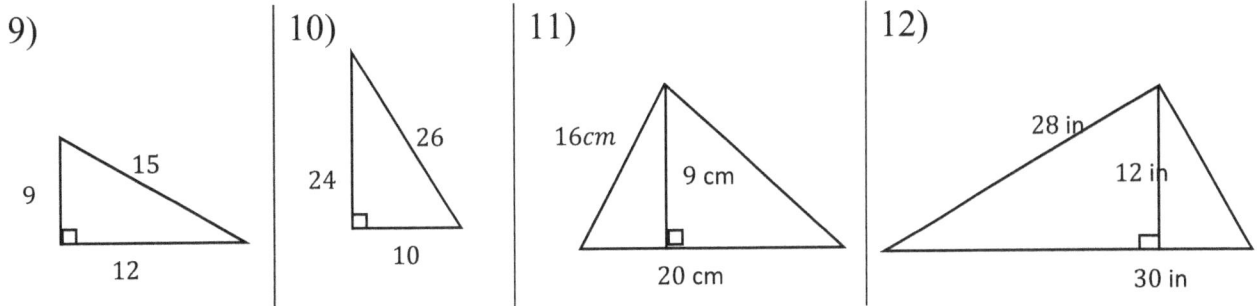

PSSA Subject Test Mathematics Grade 7

Polygons

✎ **Find the perimeter of each shape.**

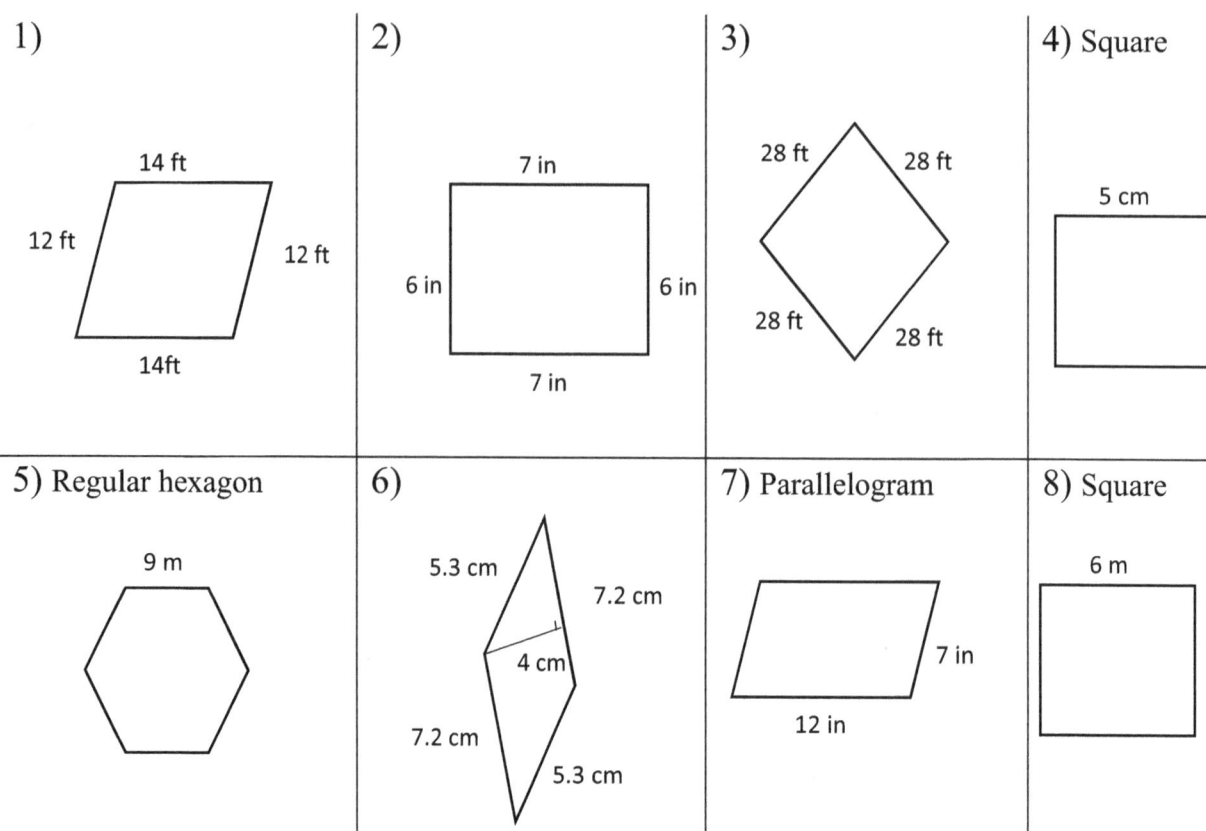

✎ **Find the area of each shape.**

9) Parallelogram
5 m, 6 m, 5 m

10) Rectangle
25 in, 12 in

11) Rectangle
16 km, 10 km

12) Square
7 in

PSSA Subject Test Mathematics Grade 7

Trapezoids

✎ Find the area of each trapezoid.

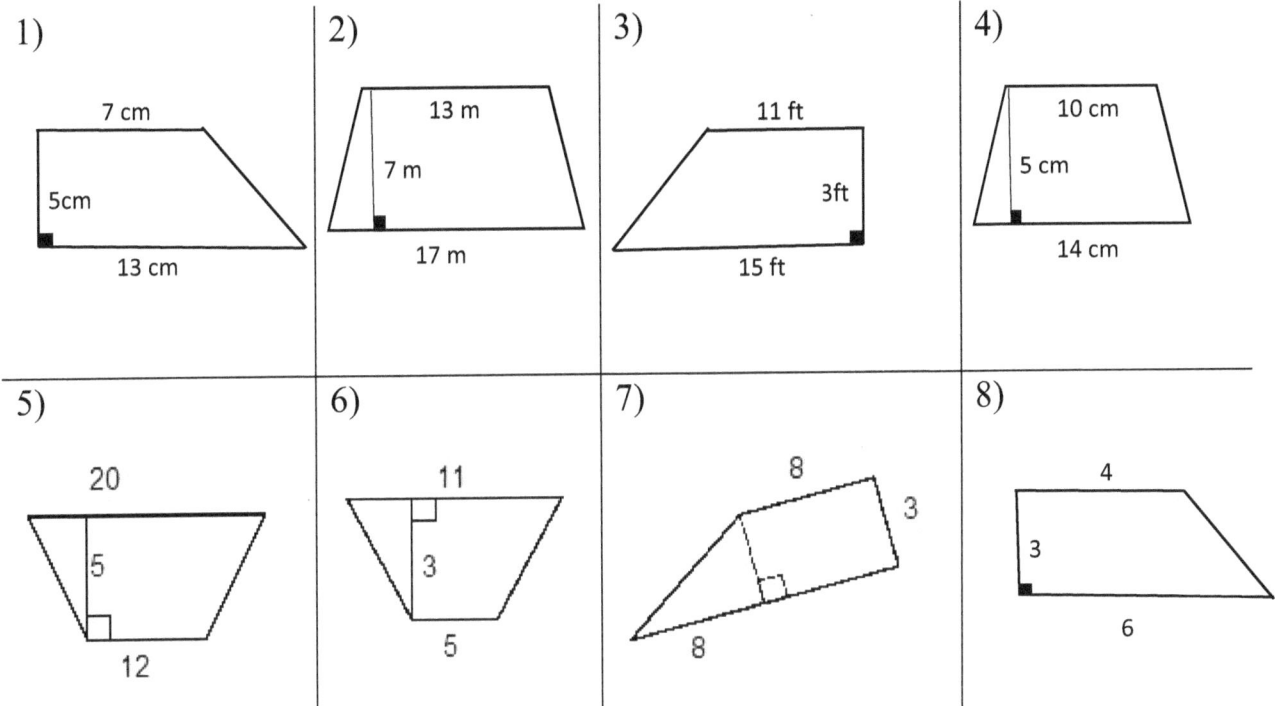

✎ Calculate.

1) A trapezoid has an area of 45 cm² and its height is 5 cm and one base is 5 cm. What is the other base length? _____

2) If a trapezoid has an area of 99 ft² and the lengths of the bases are 8 ft and 10 ft, find the height? _____

3) If a trapezoid has an area of 126 m² and its height is 14 m and one base is 6 m, find the other base length? _____

4) The area of a trapezoid is 440 ft² and its height is 22 ft. If one base of the trapezoid is 15 ft, what is the other base length? _____

Circles

✏️ **Find the area of each circle.** ($\pi = 3.14$)

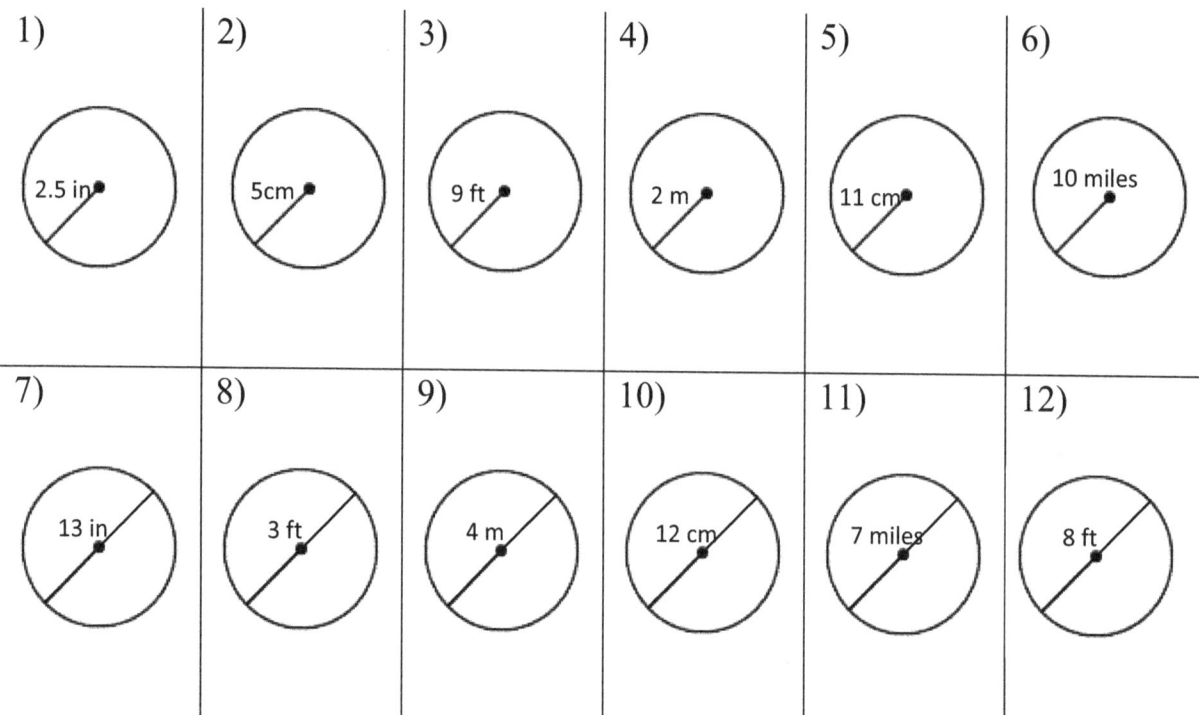

✏️ **Complete the table below.** ($\pi = 3.14$)

Circle No.	Radius	Diameter	Circumference	Area
1	1 in	2 in	6.28 in	3.14 in²
2		10 m		
3				28.26 ft²
4			47.1 mi	
5		11 km		
6	7 cm			
7		12 ft		
8				314 m²
9			56.52 in	
10	4.5 ft			

PSSA Subject Test Mathematics Grade 7

Cubes

✏️ Find the volume of each cube.

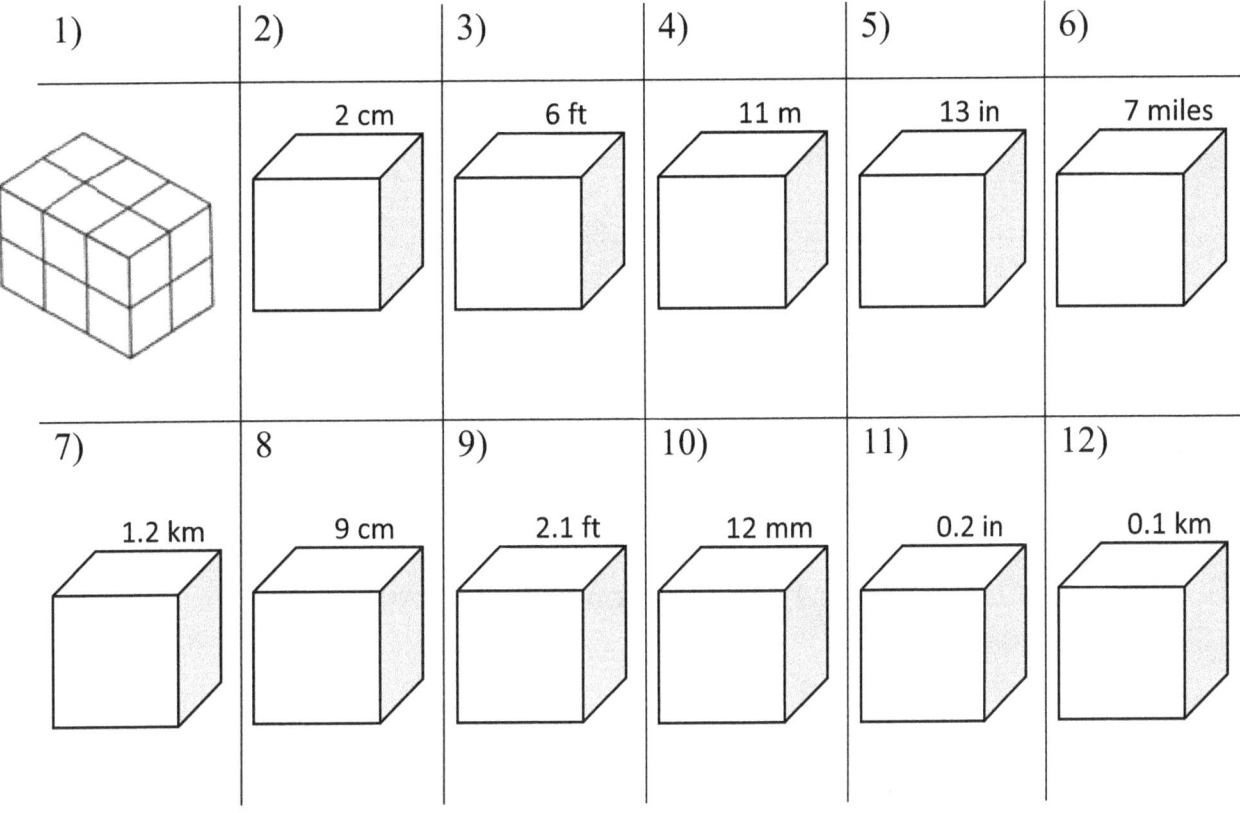

✏️ Find the surface area of each cube.

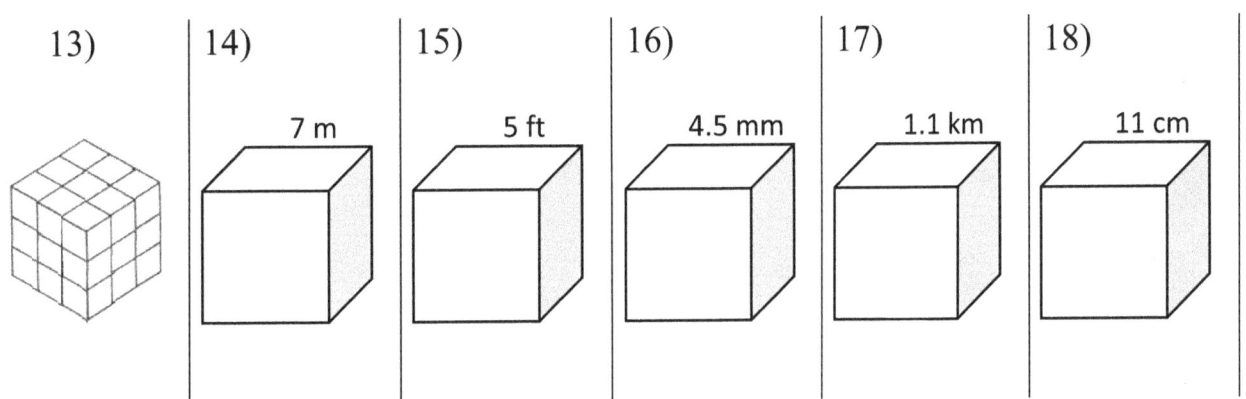

Rectangular Prism

✏️ Find the volume of each Rectangular Prism.

PSSA Subject Test Mathematics Grade 7

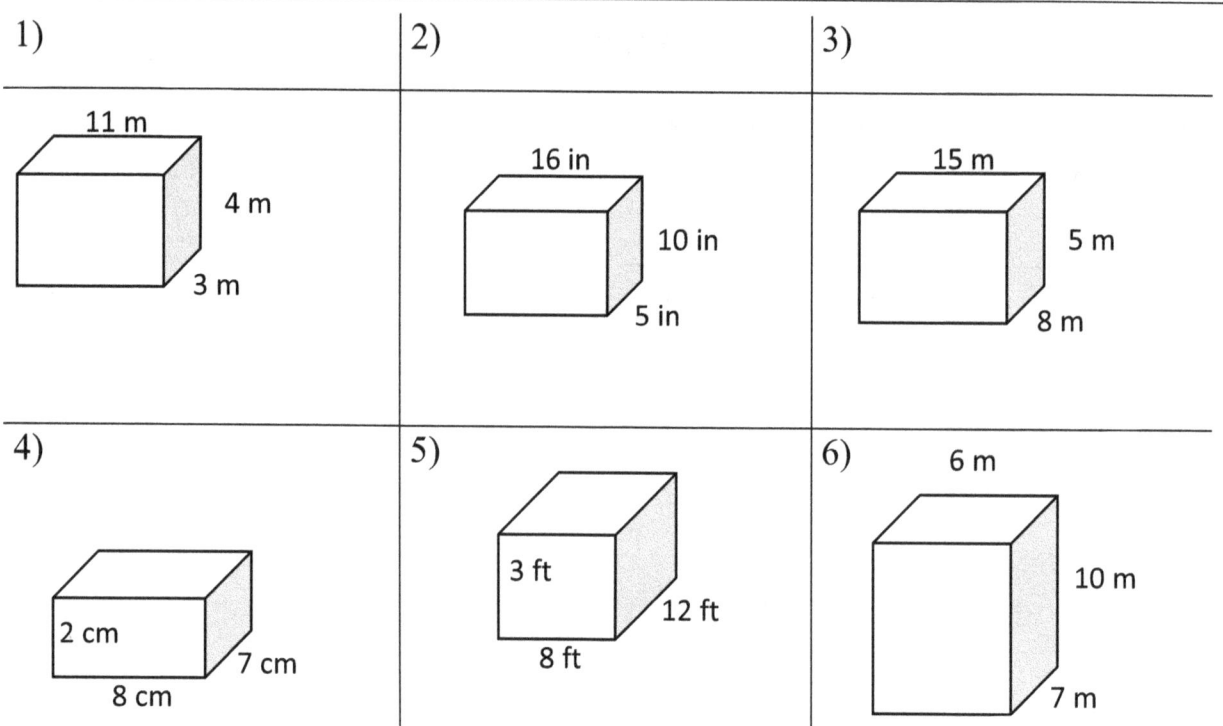

✎ Find the surface area of each Rectangular Prism.

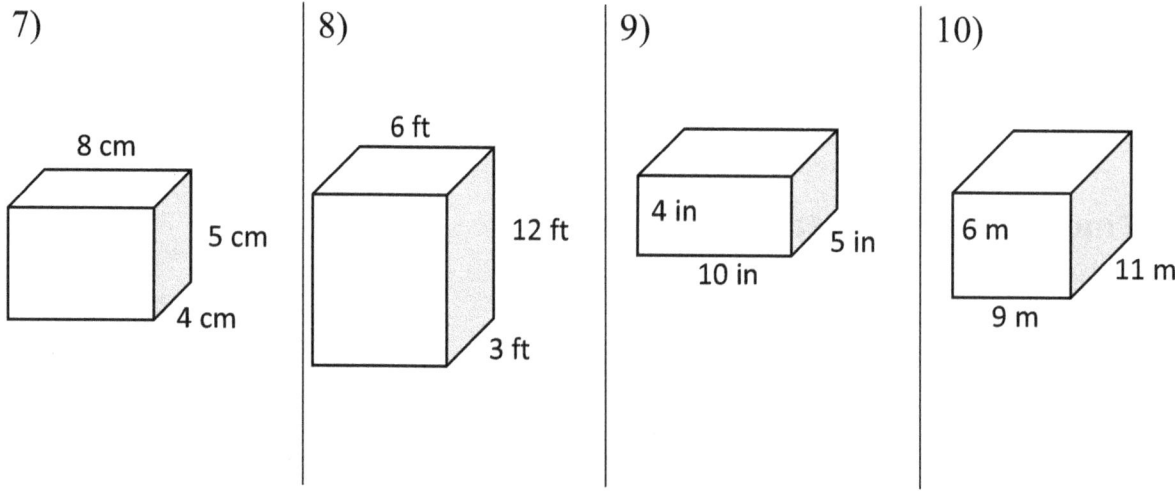

PSSA Subject Test Mathematics Grade 7

Cylinder

✏️ **Find the volume of each Cylinder. Round your answer to the nearest tenth.** ($\pi = 3.14$)

1)

2)

3)

4)

5)

6)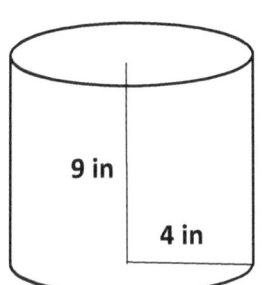

✏️ **Find the surface area of each Cylinder.** ($\pi = 3.14$)

7)

8)

9)

10)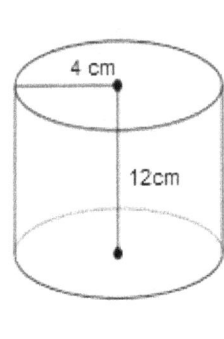

PSSA Subject Test Mathematics Grade 7

Pyramids and Cone

🖎 Find the volume of each Pyramid and Cone. ($\pi = 3.14$)

1)

2)

3)

4)

5)

6)

🖎 Find the surface area of each Pyramid and Cone. ($\pi = 3.14$)

7)

8)

9)

10)

WWW.MathNotion.Com

PSSA Subject Test Mathematics Grade 7

Answers of Worksheets

Angles

1) 16°   4) 34°   7) 90°   10) 75°
2) 96°   5) 70°   8) 75°   11) 70°
3) 59°   6) 52°   9) 33°

Pythagorean Relationship

1) No   5) Yes   9) 13   13) 15
2) Yes   6) No   10) 20   14) 30
3) No   7) Yes   11) 17   15) 36
4) Yes   8) Yes   12) 10   16) 12

Triangles

1) 60°   5) 45°   9) 54 square unites
2) 48°   6) 40°   10) 120 square unites
3) 55°   7) 45°   11) 90 square unites
4) 52°   8) 67°   12) 180 square unites

Polygons

1) 52 ft   5) 54 m   9) 30 m^2
2) 26 in   6) 25 cm   10) 300 in^2
3) 112 ft   7) 38 in   11) 160 km^2
4) 20 cm   8) 24 m   12) 49 in^2

Trapezoids

1) 50 cm^2 4) 60 cm^2 7) 36
2) 105 m^2 5) 80   8) 15
3) 39 ft^2 6) 24

Calculate

1) 13 cm   2) 11 ft   3) 12 m   4) 25 ft

Circles

1) 19.63 in^2 5) 379.94 cm^2 9) 12.56 m^2
2) 78.5 cm^2 6) 314 $miles^2$ 10) 113.04 cm^2
3) 254.34 ft^2 7) 132.67 in^2 11) 38.47 $miles^2$
4) 12.56 m^2 8) 7.07 ft^2 12) 50.24 ft^2

WWW.MathNotion.Com

PSSA Subject Test Mathematics Grade 7

Circle No.	Radius	Diameter	Circumference	Area
1	1 in	2 in	6.28 in	3.14 in^2
2	5 m	10 m	31.4 m	78.5 m^2
3	3 ft	6 ft	18.84 ft	28.26 ft^2
4	7.5 miles	15 mi	47.1 mi	176.63 mi^2
5	5.5 km	11 km	34.54 km	94.99 km^2
6	7 cm	14 cm	43.96 cm	153.86 cm^2
7	6 ft	12 ft	37.68 feet	113.04 ft^2
8	10 m	20 m	62.8 m	314 m^2
9	9 in	18 in	56.52 in	254.34 in^2
10	4.5 ft	9 ft	28.26 ft	63.585 ft^2

Cubes

1) 12
2) 8 cm^3
3) 216 ft^3
4) 1,331 m^3
5) 2,197 in^3
6) 343 $miles^3$
7) 1.728 km^3
8) 729 cm^3
9) 9.261 ft^3
10) 1,728 mm^3
11) 0.008 in^3
12) 0.001 km^3
13) 27
14) 294 m^2
15) 150 ft^2
16) 121.5 mm^2
17) 7.26 km^2
18) 726 cm^2

Rectangular Prism

1) 132 m^3
2) 800 in^3
3) 600 m^3
4) 112 cm^3
5) 288 ft^3
6) 420 m^3
7) 184 cm^2
8) 252 ft^2
9) 220 in^2
10) 438 m^2

Cylinder

1) 1,004.8 m^3
2) 214.6 cm^3
3) 9,495.4 cm^3
4) 1.1 m^3
5) 588.8 m^3
6) 452.2 in^3
7) 188.4 m^2
8) 602.9 cm^2
9) 37.7 cm^2
10) 401.9 m^2

Pyramids and Cone

1) 1,600 yd^3
2) 1,050 yd^3
3) 1,617 in^3
4) 392.5 m^3
5) 3,014.4 m^3
6) 366.33 cm^3
7) 1,440 yd^2
8) 1,536 m^2
9) 678.24 in^2
10) 1,205.76 cm^2

Chapter 11:
Statistics and Probability

Topics that you will practice in this chapter:

- ✓ Mean and Median
- ✓ Mode and Range
- ✓ Histograms
- ✓ Stem–and–Leaf Plot
- ✓ Pie Graph
- ✓ Probability Problems

Mathematics is no more computation than typing is literature.
– John Allen Paulos

> PSSA Subject Test Mathematics Grade 7

Mean and Median

✎ **Find Mean and Median of the Given Data.**

1) 8, 7, 14, 4, 8

2) 14, 8, 25, 19, 16, 33, 11

3) 23, 18, 15, 12, 17

4) 34, 14, 10, 15, 6, 11

5) 10, 19, 6, 8, 32, 20, 17

6) 17, 26, 39, 69, 20, 6

7) 40, 38, 18, 11, 9, 2, 7, 32, 41

8) 24, 21, 31, 12, 33, 32, 22

9) 16, 14, 20, 41, 15, 20, 38, 4

10) 20, 20, 30, 18, 6, 28, 12, 46

11) 12, 7, 10, 11, 16, 22

12) 10, 29, 27, 12, 2, 15, 10, 3

✎ **Calculate.**

13) In a javelin throw competition, five athletics score 56, 34, 62, 23 and 19 meters. What are their Mean and Median? _____

14) Eva went to shop and bought 8 apples, 14 peaches, 6 bananas, 4 pineapples and 12 melons. What are the Mean and Median of her purchase? _____

15) Bob has 17 black pen, 19 red pen, 14 green pens, 20 blue pens and 5 boxes of yellow pens. If the Mean and Median are 19 respectively, what is the number of yellow pens in each box? _____

PSSA Subject Test Mathematics Grade 7

Mode and Range

✎ **Find Mode and Rage of the Given Data.**

1) 4, 3, 7, 3, 3, 4
   Mode: _____   Range: _____

2) 18, 18, 24, 26, 18, 8, 14, 22
   Mode: _____   Range: _____

3) 8, 8, 8, 16, 19, 22, 20, 9, 13
   Mode: _____   Range: _____

4) 24, 24, 14, 28, 20, 18, 20, 24
   Mode: _____   Range: _____

5) 6, 21, 27, 24, 27, 27
   Mode: _____   Range: _____

6) 21, 8, 8, 7, 8, 12, 10, 22, 18, 13
   Mode: _____   Range: _____

7) 7, 4, 4, 6, 13, 13, 13, 0, 2, 2
   Mode: _____   Range: _____

8) 5, 8, 5, 14, 12, 14, 3, 5, 18
   Mode: _____   Range: _____

9) 7, 7, 7, 12, 7, 3, 8, 16, 3, 17
   Mode: _____   Range: _____

10) 15, 15, 19, 16, 4, 16, 10, 15
    Mode: _____   Range: _____

11) 6, 6, 5, 6, 42, 13, 19, 2
    Mode: _____   Range: _____

12) 8, 8, 9, 8, 9, 4, 34, 22
    Mode: _____   Range: _____

✎ **Calculate.**

13) A stationery sold 12 pencils, 56 red pens, 24 blue pens, 20 notebooks, 12 erasers, 21 rulers and 11 color pencils. What are the Mode and Range for the stationery sells?

   Mode: _____   Range: _____

14) In an English test, eight students score 10, 15, 15, 18 18, 16, 15 and 15. What are their Mode and Range? _____

15) What is the range of the first 6 even numbers greater than 8?
   _____

WWW.MathNotion.Com

PSSA Subject Test Mathematics Grade 7

Times Series

🖐 Use the following Graph to complete the table.

Day	Distance (km)
1	
2	

The following table shows the number of births in the US from 2007 to 2012 (in millions).

Year	Number of births (in millions)
2007	4.15
2008	3.70
2009	3.45
2010	3.20
2011	1.75
2012	2.98

Draw a Time Series for the table.

Stem-and-Leaf Plot

✎ **Make stem ad leaf plots for the given data.**

1) 24, 26, 29, 20, 53, 27, 51, 55, 36, 21, 37, 30

   Stem | Leaf plot

2) 11, 59, 66, 14, 18, 19, 59, 65, 69, 61, 68, 65

   Stem | Leaf plot

3) 121, 55, 66, 54, 112, 128, 63, 125, 59, 123, 68, 119

   Stem | Leaf plot

4) 51, 32, 100, 56, 84, 36, 107, 56, 85, 39, 56, 106, 89

   Stem | Leaf plot

5) 33, 89, 19, 87, 81, 16, 11, 30, 86, 35, 17, 35, 13

   Stem | Leaf plot

6) 60, 92, 22, 25, 67, 93, 95, 62, 21, 64, 98, 29

   Stem | Leaf plot

Pie Graph

The circle graph below shows all Robert's expenses for last month. Robert spent $140 on his hobbies last month.

Answer following questions based on the Pie graph.

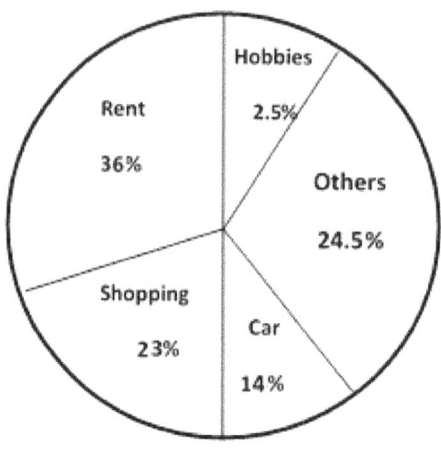

1) How much was Robert's total expenses last month? _____

2) How much did Robert spend on his car last month? _____

3) How much did Robert spend for shopping last month? _____

4) How much did Robert spend on his rent last month? _____

5) What fraction is Robert's expenses for his rent and car out of his total expenses last month? _____

PSSA Subject Test Mathematics Grade 7

Probability Problems

✏️ **Calculate.**

1) A number is chosen at random from 1 to 10. Find the probability of selecting number 6 or smaller numbers. _____

2) Bag A contains 18 red marbles and 6 green marbles. Bag B contains 16 black marbles and 8 orange marbles. What is the probability of selecting a green marble at random from bag A? What is the probability of selecting a black marble at random from Bag B? _____

3) A number is chosen at random from 1 to 20. What is the probability of selecting multiples of 4? _____

4) A card is chosen from a well-shuffled deck of 52 cards. What is the probability that the card will be a queen? _____

5) A number is chosen at random from 1 to 15. What is the probability of selecting a multiple of 3 or 5? _____

A spinner numbered 1–8, is spun once. What is the probability of spinning …?

6) an Odd number? _____   7) a multiple of 2? _____

8) a multiple of 5? _____   9) number 10? _____

PSSA Subject Test Mathematics Grade 7

Answers of Worksheets

Mean and Median

1) Mean: 8.2, Median: 8
2) Mean: 18, Median: 16
3) Mean: 17, Median: 17
4) Mean: 15, Median: 12.5
5) Mean: 16, Median: 17
6) Mean: 29.5, Median: 23
7) Mean: 22, Median: 18
8) Mean: 25, Median: 24
9) Mean: 21, Median: 18
10) Mean: 22.5, Median: 20
11) Mean: 13, Median: 11.5
12) Mean: 13.5, Median: 11
13) Mean: 38.8, Median: 34
14) Mean: 8.8, Median: 8
15) 5

Mode and Range

1) Mode: 3, Range: 4
2) Mode: 18, Range: 18
3) Mode: 8, Range: 14
4) Mode: 24, Range: 14
5) Mode: 27, Range: 21
6) Mode: 8, Range: 15
7) Mode: 13, Range: 13
8) Mode: 5, Range: 15
9) Mode: 7, Range: 14
10) Mode: 15, Range: 15
11) Mode: 6, Range: 40
12) Mode: 8, Range: 30
13) Mode: 12, Range: 45
14) Mode: 15, Range: 8
15) 10

Time series

Day	Distance (km)
1	335
2	496
3	270
4	610
5	320
6	400

Stem–And–Leaf Plot

1)

Stem	leaf
2	0 1 4 6 7 9
3	0 6 7
5	1 3 5

2)

Stem	leaf
1	1 4 8 9
5	9 9
6	1 5 5 6 8 9

3)

Stem	leaf
5	4 5 9
6	3 6 8
11	2 9
12	1 3 5 8

WWW.MathNotion.Com

PSSA Subject Test Mathematics Grade 7

4)

Stem	leaf
3	2 6 9
5	1 6 6 6
8	4 5 9
10	0 6 7

5)

Stem	leaf
1	1 3 6 7 9
3	0 3 5 5
8	1 6 7 9

6)

Stem	leaf
2	2 1 5 9
6	0 2 4 7
9	2 3 5 8

Pie Graph

1) $5,600
2) $784
3) $1,288
4) $2,016
5) $\frac{1}{2}$

Probability Problems

1) $\frac{3}{5}$
2) $\frac{1}{4}, \frac{2}{3}$
3) $\frac{1}{4}$
4) $\frac{1}{13}$
5) $\frac{7}{15}$
6) $\frac{1}{2}$
7) $\frac{1}{2}$
8) $\frac{1}{8}$
9) 0

WWW.MathNotion.Com

PSSA Subject Test Mathematics Grade 7

PSSA Subject Test Mathematics Grade 7

Chapter 12 : PSSA Math Practice Tests

Time to Test

Time to refine your skill with a practice examination.

Take two practice Grade 7 PSSA Math Tests to simulate the test day experience.

After you've finished, score your test using the answer key.

Before You Start

- You'll need a pencil and scratch papers to take the test.

- For this practice test, don't time yourself. Spend time as much as you need.

- It's okay to guess. You won't lose any points if you're wrong.

- After you've finished the test, review the answer key to see where you went go.

Calculators are not permitted for students taking the

PSSA Mathematics Grade 7.

Good Luck!

PSSA Subject Test Mathematics Grade 7

Grade 7 PSSA Mathematics Formula Sheet

Formulas that you may need to work questions on this test are found below. You may refer to this page at any time during the mathematics test.

Triangle

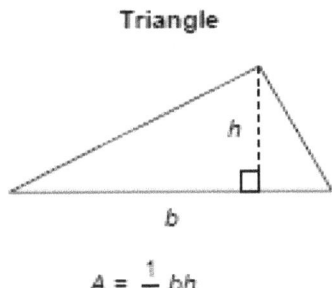

$A = \dfrac{1}{2} bh$

Rectangle

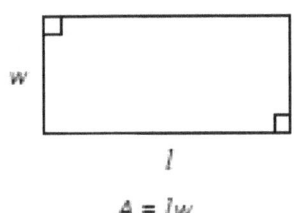

$A = lw$

Square

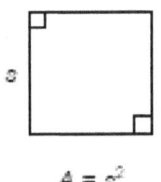

$A = s^2$

Parallelogram

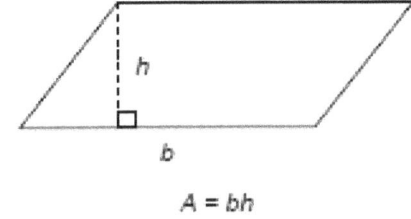

$A = bh$

Trapezoid

$A = \dfrac{1}{2} h(b_1 + b_2)$

Rectangular Prism

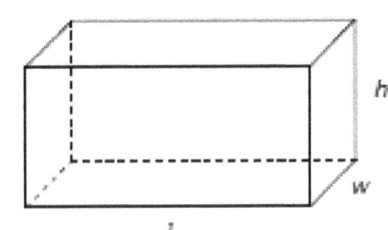

$V = lwh \qquad SA = 2lw + 2lh + 2wh$

Cube

$V = s \cdot s \cdot s \qquad SA = 6s^2$

Triangular Prism

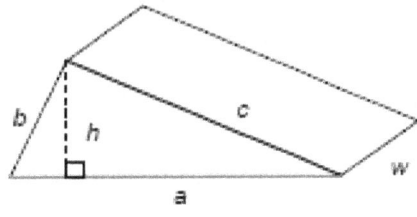

$SA = ah + aw + bw + cw$

WWW.MathNotion.Com

The Pennsylvania System of School Assessment

PSSA Practice Test 1

Mathematics

GRADE 7

- 20 Questions
- Calculators are permitted for this practice test.

Pennsylvania Department of Education Bureau of Curriculum, Assessment, and Instruction - *Month Year*

PSSA Subject Test Mathematics Grade 7

1) The price of a car was $18,000 in 2014, $13,500 in 2015 and $10,125 in 2016.

   What is the rate of depreciation of the price of car per year?

   A. 10 %

   B. 35 %

   C. 25%

   D. 15 %

2) What is the value of x in the following equation?

   $$\frac{3}{5}x + \frac{1}{6} = \frac{1}{3}$$

   A. $\frac{3}{8}$

   B. $\frac{5}{18}$

   C. $\frac{5}{12}$

   D. $\frac{1}{30}$

3) What is the surface area of the cylinder below?

   A. 176 π

   B. 280 π

   C. 352 π

   D. 480 π

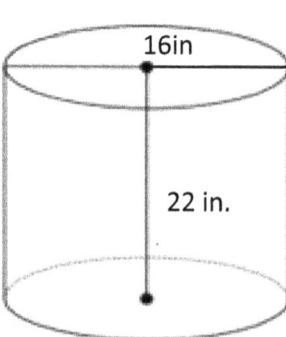

PSSA Subject Test Mathematics Grade 7

4) Anita's trick or treat bag contains 15 pieces of chocolate, 25 suckers, 26 pieces of gum, 44 pieces of licorice. If she randomly pulls a piece of candy from her bag, what is the probability of her pulling out a piece of sucker?

A. $\frac{2}{15}$

B. $\frac{1}{12}$

C. $\frac{3}{22}$

D. $\frac{5}{22}$

5) Which of the following shows the numbers in descending order?

$$\frac{6}{5}, 0.67, 79\%, \frac{3}{8}$$

A. $79\%, 0.67, \frac{6}{5}, \frac{3}{8}$

B. $79\%, 0.67, \frac{3}{8}, \frac{6}{5}$

C. $\frac{3}{8}, 0.67, 79\%, \frac{6}{5}$

D. $\frac{6}{5}, 79\%, 0.67, \frac{3}{8}$

6) Which of the following points lies on the line $2x + 5y = 16$?

A. (3, 5)

B. (–5, 3)

C. (–2, 4)

D. (5, –3)

WWW.MathNotion.Com

PSSA Subject Test Mathematics Grade 7

7) The mean of 42 test scores was calculated as 65. But it turned out that one of the scores was misread as 63 but it was 36. What is the correct mean of the data?

Write your answer in the box below.

8) The width of a box is one fifth of its length. The height of the box is one fifth of its width. If the length of the box is 50 cm, what is the volume of the box?

   A. 2,000 cm³

   B. 1,250 cm³

   C. 1,500 cm³

   D. 1,000 cm³

9) What is the slope of a line that is parallel to the line

   $-20x - 5y = 35$?

   Write your answer in the box below.

10) Which of the following graphs represents the compound inequality

    $-1 \leq 5x - 6 < 4$?

    A.

    B.

    C.

    D.

PSSA Subject Test Mathematics Grade 7

11) A bank is offering 2.8% simple interest on a savings account. If you deposit $15,000, how much interest will you earn in four years?

   A. $340

   B. $1,680

   C. $2,880

   D. $4,260

12) What is the volume of a box with the following dimensions?

   Height = 9 cm   Width = 5 cm   Length = 6 cm

   A. 30 cm^3

   B. 45 cm^3

   C. 270 cm^3

   D. 540 cm^3

13) A shirt costing $820 is discounted 30%. After a month, the shirt is discounted another 30%. Which of the following expressions can be used to find the selling price of the shirt?

   A. (820) (0.30)

   B. (820) − 820 (0.30)

   C. (820) (0.30) − (820) (0.70)

   D. (820) (0.70) (0.70)

14) Which graph corresponds to the following inequalities?

$$y \leq -3x + 6$$

$$-4x + 2y \leq 8$$

A.

B.

C.

D.
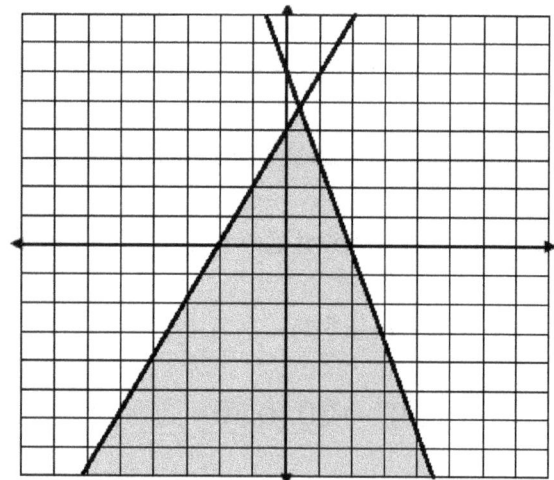

15) Mr. Carlos family are choosing a menu for their reception. They have 3 choices of appetizers, 7 choices of entrees, 8 choices of cake. How many different menu combinations are possible for them to choose?

A. 36

B. 56

C. 168

D. 280

16) What is the area of the shaded region?

Write your answer in the box below.

17) When a number is subtracted from 60 and the difference is divided by that number, the result is 4. What is the value of the number?

A. 14

B. 24

C. 16

D. 12

PSSA Subject Test Mathematics Grade 7

18) In two successive years, the population of a town is increased by 10% and 35%.

What percent of its population is increased after two years?

A. 62.5

B. 48.5

C. 42.6

D. 20.4

19) What is the value of the expression $5(2x + y) - (10 - 3x)^2$ when $x = 2$ and y= −2 ?

Write your answer in the box below.

20) Two dice are thrown simultaneously, what is the probability of getting a sum of 5 or 11?

A. $\frac{7}{18}$

B. $\frac{2}{9}$

C. $\frac{1}{8}$

D. $\frac{1}{6}$

Practice Test 1
This is the End of this Section.

PSSA Subject Test Mathematics Grade 7

The Pennsylvania System of School Assessment

PSSA Practice Test 2

Mathematics

GRADE 7

❖ 20 Questions
❖ Calculators are permitted for this practice test.

Pennsylvania Department of Education Bureau of Curriculum, Assessment, and Instruction - *Month Year*

PSSA Subject Test Mathematics Grade 7

1) If $x = -7$, which equation is true?

   A. $x(4x + 5) = -90$

   B. $8(4 - x) = 88$

   C. $2(8x + 3) = -75$

   D. $11x - 7 = -80$

2) In a bag of small balls $\frac{1}{6}$ are black, $\frac{1}{4}$ are white, $\frac{1}{12}$ are red and the remaining 8 blue. How many balls are white?

   A. 12

   B. 8

   C. 4

   D. 18

3) A boat sails 12 miles south and then 16 miles east. How far is the boat from its start point?

   Write your answer in the box below.

4) Simplify $5y^2(4x^5y)^4 =$

   Write your answer in the box below.

WWW.MathNotion.Com

PSSA Subject Test Mathematics Grade 7

5) Sophia purchased a sofa for $612.50. The sofa is regularly priced at $625. What was the percent discount Sophia received on the sofa?

   A. 12%

   B. 2%

   C. 98%

   D. 88%

6) The score of Elise was one eighth as that of Mia and the score of Stella was twice that of Mia. If the score of Stella was 80, what is the score of Elise?

   A. 8

   B. 40

   C. 5

   D. 15

7) A bag contains 16 balls: six green, four black, one blue, one brown, three red and one white. If 12 balls are removed from the bag at random, what is the probability that a brown ball has been removed?

   A. $\frac{3}{4}$

   B. $\frac{1}{12}$

   C. $\frac{5}{16}$

   D. $\frac{1}{16}$

PSSA Subject Test Mathematics Grade 7

8) What is the area of the shaded region?

   A. 64π in²

   B. 96π in²

   C. 105π in²

   D. 169π in²

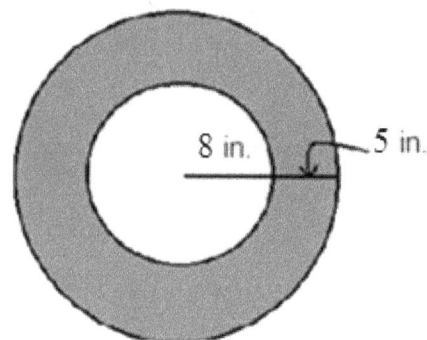

9) A pizza cut into 12 parts. Elise and her sister Etta ordered two pizzas. Elise ate $\frac{5}{6}$ of her pizza and Etta ate $\frac{5}{12}$ of her pizza. What part of the two pizzas was left?

   A. $\frac{3}{8}$

   B. $\frac{3}{16}$

   C. $\frac{5}{12}$

   D. $\frac{1}{3}$

10) The marked price of a computer is D dollar. Its price decreased by 35% in January and later increased by 20% in February. What is the final price of the computer in D dollar?

   A. 0.78 D

   B. 0.68 D

   C. 0.88 D

   D. 1.08 D

PSSA Subject Test Mathematics Grade 7

11) [6 × (−16) + 5] − (5) + [8 × 9] ÷ 4 =?

Write your answer in the box below.

12) From last year, the price of gasoline has increased from $1.75 per gallon to $2.454 per gallon. The new price is what percent of the original price?

A. 60 %

B. 40 %

C. 140 %

D. 130 %

13) What is the median of these numbers? 11, 36, 39, 3, 52, 33, 25

Write your answer in the box below.

14) The following trapezoid are similar. What is the value of x ?

A. 5

B. 3

C. 25

D. 12

 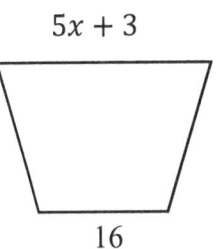

WWW.MathNotion.Com

15) Three times the price of a laptop is equal to five times the price of a computer. If the price of laptop is $400 more than the computer, what is the price of the computer?

   A. 450

   B. 400

   C. 650

   D. 600

16) What is the volume of the following square pyramid?

   A. 456 m³

   B. 664 m³

   C. 768 m³

   D. 1,220 m³

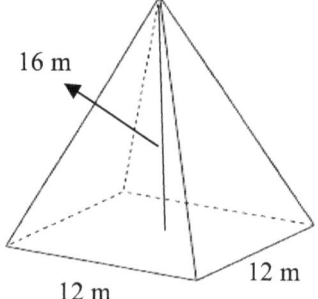

17) A bank is offering 2.6% simple interest on a savings account. If you deposit $30,000, how much interest will you earn in five years?

   A. $3,900

   B. $900

   C. $2,600

   D. $4,800

PSSA Subject Test Mathematics Grade 7

18) Which of the following points lies on the line $8x + 5y = 7$?

   A. (5, 3)

   B. (–1, 3)

   C. (–3, 4)

   D. (–2, –3)

19) An angle is equal to one third of its supplement. What is the measure of that angle?

   Write your answer in the box below.

   ☐

20) John traveled 630 km in 9 hours and Alice traveled 360 km in 4 hours. What is the ratio of the average speed of John to average speed of Alice?

   A. 7: 8

   B. 3: 7

   C. 7: 9

   D. 8: 9

Practice Test 2

This is the End of this Section.

PSSA Subject Test Mathematics Grade 7

Chapter 13 : Answers and Explanations

PSSA Practice Tests
Answer Key

❋ Now, it's time to review your results to see where you went wrong and what areas you need to improve!

PSSA - Mathematics

Practice Test - 1

1	C	11	B
2	B	12	C
3	D	13	D
4	D	14	D
5	D	15	C
6	C	16	$179ft^2$
7	64.36	17	D
8	D	18	B
9	−4	19	−6
10	A	20	D

Practice Test - 2

1	B	11	−78
2	C	12	C
3	20	13	33
4	$1,280x^{20}y^6$	14	A
5	B	15	D
6	C	16	C
7	D	17	A
8	C	18	B
9	A	19	45
10	A	20	C

Practice Test 1
Answers and Explanations

1) Answer: C

Use this formula: Percent of Change = $\frac{New\ Value - Old\ Value}{Old\ Value} \times 100\ \%$

$\frac{13,500-18,000}{18,000} \times 100\ \% = -25\ \%$ and $\frac{10,125-13,500}{13,500} \times 100\ \% = -25\%$

2) Answer: B

$\frac{3}{5}x + \frac{1}{6} = \frac{1}{3} \Rightarrow \frac{3}{5}x = \frac{1}{6} \Rightarrow x = \frac{5}{3} \times \frac{1}{6} \Rightarrow x = \frac{5}{18}$

3) Answer: D

Surface Area of a cylinder = $2\pi r\ (r + h)$,

The radius of the cylinder is 8 (16 ÷ 2) inches, and its height is 22 inches. Therefore,

Surface Area of a cylinder = $2\pi\ (8)\ (8 + 22) = 480\ \pi$

4) Answer: D

Probability = $\frac{number\ of\ desired\ outcomes}{number\ of\ total\ outcomes} = \frac{25}{15+25+26+44} = \frac{25}{110} = \frac{5}{22}$

5) Answer: D

Change the numbers to decimal and then compare.

$\frac{3}{8} = 0.375;\ 0.67;\ 79\% = 0.79;\ \frac{6}{5} = 1.2$

Therefore $\frac{6}{5} > 79\% > 0.67 > \frac{3}{8}$

6) Answer: C

Input $(-2, 4)$ in the $2x + 5y = 16$ formula instead of x and y. So, we have: $2(-2) + 5(4)$
$= 16 \rightarrow -4 + 20 = 16$

7) Answer: 64.36

average (mean) = $\frac{sum\ of\ terms}{number\ of\ terms} \Rightarrow 65 = \frac{sum\ of\ terms}{42} \Rightarrow sum = 65 \times 42 = 2,730$

The difference of 63 and 36 is 27. Therefore, 27 should be subtracted from the sum.

$2,730 - 27 = 2,703$

mean = $\frac{sum\ of\ terms}{number\ of\ terms} \Rightarrow mean = \frac{2,703}{42} = 64.36$

PSSA Subject Test Mathematics Grade 7

8) Answer: D

If the length of the box is 50, then the width of the box is one fifth of it, 10, and the height of the box is 2 (one fifth of the width). The volume of the box is:

$V = l \times w \times h = (50) \times (10) \times (2) = 1{,}000$

9) Answer: −4

The equation of a line in slope intercept form is: $y = mx + b$

Solve for y.

$-20x - 5y = 35 \Rightarrow -5y = 35 + 20x \Rightarrow y = (35 + 20x) \div (-5) \Rightarrow$
$y = -4x - 7$

The slope of this line is -4. Parallel lines have same slopes.

10) Answer: A

Solve for x.

$-1 \leq 5x - 6 < 4 \Rightarrow$ (add 6 all sides) $-1 + 6 \leq 5x - 6 + 6 < 4 + 6 \Rightarrow$
$5 \leq 5x < 10 \Rightarrow$ (divide all sides by 5) $1 \leq x < 2$

x is between 1 and 2. Choice A represent this inequality.

11) Answer: B

Use simple interest formula: $I = prt$ (I = interest, p = principal, r = rate, t = time)

$I = (15{,}000)(0.028)(4) = 1{,}680$

12) Answer: C

Volume of a box = length × width × height = $9 \times 5 \times 6 = 270$

13) Answer: D

To find the discount, multiply the number by (100% − rate of discount).

Therefore, for the first discount we get: $(820)(100\% - 30\%) = (820)(0.70)$

For the next 30 % discount: $(820)(0.70)(0.70)$

14) Answer: D.

For each option, choose a point in the solution part and check it on both inequalities.

$$y \leq -3x + 6$$
$$-4x + 2y \leq 8$$

WWW.MathNotion.Com

PSSA Subject Test Mathematics Grade 7

A. Point (1, 7) is in the solution section. Let's check the point in both inequalities.

$7 \leq -3(1) + 6$, That's not true

B. Let's choose this point (–3, 2); $2 \leq -3(-3) + 6$, That's true!

$-4(-3) + 2(2) \leq 8 \Rightarrow 14 \leq 8$, That's not true

C. Let's choose this point (3, 3); $3 \leq -3(3) + 6$, That's not true

D. Let's choose this point (0, 0); $0 \leq -3(0) + 6$, That's true!

$-4(0) + 2(0) \leq 8 \Rightarrow 0 \leq 8$, it works (this point works in both)

15) **Answer: C**

To find the number of possible outfit combinations, multiply number of options for each factor: $3 \times 7 \times 8 = 168$

16) **Answer:** $179 ft^2$

Use the area of rectangle formula ($s = a \times b$).

To find area of the shaded region subtract the smaller rectangle from bigger rectangle.

$S_1 - S_2 = (18ft \times 15ft) - (13ft \times 7ft) \Rightarrow S_1 - S_2 = 270ft^2 - 91ft^2 = 179ft^2$

17) **Answer: D**

Let the number be x. Then:

$\frac{60-x}{x} = 4 \rightarrow 4x = 60 - x \rightarrow 5x = 60 \rightarrow x = 12$

18) **Answer: B**

The population is increased by 10% and 35%. 10% increase changes the population to 100% of original population.

For the second increase, multiply the result by 135%.

$(1.10) \times (1.35) = 1.485 = 148.5\%$

48.5 percent of the population is increased after two years.

19) **Answer:** -6

When $x = 2$ and $y = -2$,

Substitute: $5(2x + y) - (10 - 3x)^2 = 5(2(2) + (-2)) - (10 - 3(2))^2$

$= 5(4 - 2) - (10 - 6)^2 = 10 - 16 = -6$

20) **Answer: D**

The options to get sum of 5: (1 & 5) and (5 & 1), (2 & 3) and (3 & 2)

so, we have 4 options.

The options to get sum of 11: (5 & 6) and (6 & 5) we have 3 options.

To get the sum of 5 or 11 for two dice, we have 6 options: $4 + 2 = 6$

Since, we have $6 \times 6 = 36$ total options, the probability of getting a sum of 5 and 11 is 6 out of 36 or $\frac{6}{36} = \frac{1}{6}$.

PSSA Subject Test Mathematics Grade 7

Practice Test 2
Answers and Explanations

1) Answer: B

Only option B is correct. Other options don't work in the equation.

$8(4-(-7)) = 88$

2) Answer: C

Let x be the number of balls. Then:

$\frac{1}{6}x + \frac{1}{4}x + \frac{1}{12}x + 8 = x$

$(\frac{1}{6} + \frac{1}{4} + \frac{1}{12})x + 8 = x$

$(\frac{1}{2})x + 8 = x \Rightarrow x = 16$

In the bag of small balls $\frac{1}{4}$ are white, then: $\frac{16}{4} = 4$

There are 4 white balls in the bag.

3) Answer: 20.

Use the information provided in the question to draw the shape.

Use Pythagorean Theorem: $a^2 + b^2 = c^2$

$12^2 + 16^2 = c^2 \Rightarrow 144 + 256 = c^2 \Rightarrow 400 = c^2$

$\Rightarrow c = 20$

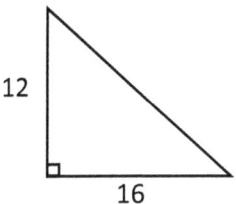

4) Answer: $1,280x^{20}y^6$

Simplify: $5y^2(4x^5y)^4 = 5y^2(256x^{20}y^4) = 1,280x^{20}y^6$

5) Answer: B

The question is this: 612.50 is what percent of 625?

Use percent formula: part $= \frac{\text{percent}}{100} \times$ whole

$612.50 = \frac{\text{percent}}{100} \times 625 \Rightarrow 612.50 = \frac{\text{percent} \times 625}{100}$

$\Rightarrow 61,250 =$ percent $\times 625 \Rightarrow$ percent $= \frac{61,250}{625} = 98$

612.50 is 98 % of 625. Therefore, the discount is: 100% − 98% = 2%

PSSA Subject Test Mathematics Grade 7

6) Answer: C

If the score of Stella was 80, then the score of Mia is 40. Since the score of Elise was one eighth as that of Mia, therefore, the score of Elise is 5.

7) Answer: D

If 12 balls are removed from the bag at random, there will be one ball in the bag. The probability of choosing one brown ball is 1 out of 16. Therefore, the probability of not choosing one brown ball is 12 out of 16 and the probability of having not brown balls after removing 12 balls is the same.

8) Answer: C

To find the area of the shaded region, find the difference of the area of two circles. (S_1: the area of bigger circle. S_2: the area of the smaller circle)

Use the area of circle formula. $S = \pi r^2$

$S1 - S2 = \pi(8+5)^2 - \pi(8in)^2 \Rightarrow S1 - S2 = \pi 169 in^2 - \pi 64\ in^2 \Rightarrow S1 - S2 = 105\pi\ in^2$

9) Answer: A

Elise ate $\frac{5}{6}$ of 12 parts of his pizza that it means 10 parts out of 12 parts ($\frac{5}{6}$ of 12 parts $= x \Rightarrow x = 10$) and left 2 parts.

Etta ate $\frac{5}{12}$ of 12 parts of her pizza that it means 5 part out of 12 parts ($\frac{5}{12}$ of 12 parts $= x \Rightarrow x = 5$) and left 7 parts.

Therefore, they ate (10 + 5) parts out of (12 + 12) parts of their pizza and left (2 + 7) parts out of (12 + 12) parts of their pizza. It means: $\frac{9}{24}$

After simplification we have: $\frac{3}{8}$

10) Answer: A

To find the discount, multiply the number by (100% – rate of discount).

Therefore, for the first discount we get: (D) (100% – 35%) = (D) (0.65) = 0.65 D

For increase of 20 %: (0.65 D) (100% + 20%) = (0.65 D) (1.20) = 0.78 D = 78% of D

PSSA Subject Test Mathematics Grade 7

11) Answer: −78

Use PEMDAS (order of operation): $[6 \times (-16) + 5] - (5) + [8 \times 9] \div 4 =$

$= [-96 + 5] - 5 + [72] \div 4 = [-91] - 5 + 18 = -78$

12) Answer: C

The question is this: 2.45 is what percent of 1.75?

Use percent formula: part $= \frac{percent}{100} \times$ whole $= \frac{percent}{100} \times 1.75$

$\Rightarrow 2.45 = \frac{percent \times 1.75}{100} \Rightarrow 245 =$ percent $\times 1.75 \Rightarrow$ percent $= \frac{245}{1.75} = 140$

13) Answer: 33

Write the numbers in order: 3, 11, 25, 33, 36, 39, 52.

Since we have 7 numbers (7 is odd), then the median is the number in the middle, which is 33.

14) Answer: A

Write the ratio and solve for x.

$\frac{70}{40} = \frac{5x+3}{16} \Rightarrow 40(5x + 3) = 70 \times 16 \Rightarrow 5x + 3 = 1{,}120 \div 40 \Rightarrow 5x + 3 = 28 \Rightarrow x = 5$

15) Answer: D

Let L be the price of laptop and C be the price of computer.

$3(L) = 5(C)$ and $L = \$400 + C$

Therefore, $3(\$400 + C) = 5C \Rightarrow \$1{,}200 + 3C = 5C \Rightarrow C = \$600$

16) Answer: C

Use the volume of square pyramid formula.

$V = \frac{1}{3} a^2 h \Rightarrow V = \frac{1}{3}(12m)^2 \times 16m \Rightarrow V = 768$ m³

17) Answer: A

Use simple interest formula: $I = prt$ (I = interest, p = principal, r = rate, t = time)

$I = (30{,}000)(0.026)(5) = 3{,}900$

18) Answer: B

Input the points instead of x and y in the formula. Only option B works in the equation.

$8x + 5y = 7 \Rightarrow 8(-1) + 5(3) = 7$

PSSA Subject Test Mathematics Grade 7

19) Answer: 45

The sum of supplement angles is 180. Let x be that angle. Therefore,

$x + 3x = 180$

$4x = 180$, divide both sides by 4: $x = 45$

20) Answer: C

The average speed of John is: $630 \div 9 = 70$ km

The average speed of Alice is: $360 \div 4 = 90$ km

Write the ratio and simplify: $70: 90 \Rightarrow 7: 9$.

"End"

www.ingramcontent.com/pod-product-compliance
Lightning Source LLC
Chambersburg PA
CBHW080442110426
42743CB00016B/3246